I0473940

NISTIR 7772

Information Model for Disassembly for Reuse, Recycle, and Remanufacturing

Shaw C. Feng[1]
Hanmin Lee[1,4]
Che B. Joung[1]
Thomas Kramer[2]
Parisa Ghodous[5]
Ram D. Sriram[3]

[1] *Manufacturing Systems Integration Division*
[2] *Intelligent Systems Division*
Engineering Laboratory

[3] *Software and Systems Division*
Information Technology Laboratory
NIST

[4] *Korea Institute of Machinery and Materials*
South Korea

[5] *University of Claude Bernard Lyon I*
France

February 2011

U.S. Department of Commerce
Gary Locke, Secretary

National Institute of Standards and Technology
Patrick D. Gallagher, Director

Table of Contents

2

Abstract

Disassembly process is an essential activity in the mid and end of a product's service life cycle. Maintenance is a mid product life activity, which involves disassembly, cleaning, repair, replace, inspection, and reassembly. Remanufacture, reuse, and recycle are activities at the end of a product's service life. The main purpose of performing these activities is to increase sustainability, which includes reduce environmental impact, increase economic, and extend societal benefits to meet the goal of sustainable development. This report describes a developed disassembly information model. It is an integrated information model with the following key components: feature, tolerance, workpiece, material content, equipment, and process. The model supports the modeling of disassembly features, sequence, operations, decision making, and workflow. Case study using real products and their specific disassembly processes is provided to verify the developed model.

Key words:

design for disassembly, disassembly for remanufacturing, disassembly for reuse, disassembly modeling, feature-based disassembly, product disassembly.

1. Introduction

Manufacturing is the fundamental support of an industrialized society. Two major challenges faced by manufacturing industries today are global natural resource depletion and environmental pollution. For example, the use of copper has grown exponentially worldwide. Refined copper demand has grown from 0.5 million metric tons at the beginning of the 20th century to 17.3 million tons in 2006. Figure 1 shows the annual world copper usage from 1900 to 2006. In 2009, the estimated world copper usage is 20.5 million tons, according to two studies by the International Copper Study Group [ICSG 2006-1, ICSG 2006-2].

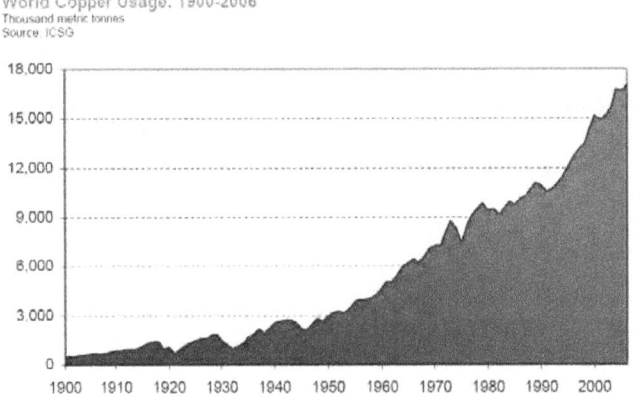

Figure 1. Annual world copper usage

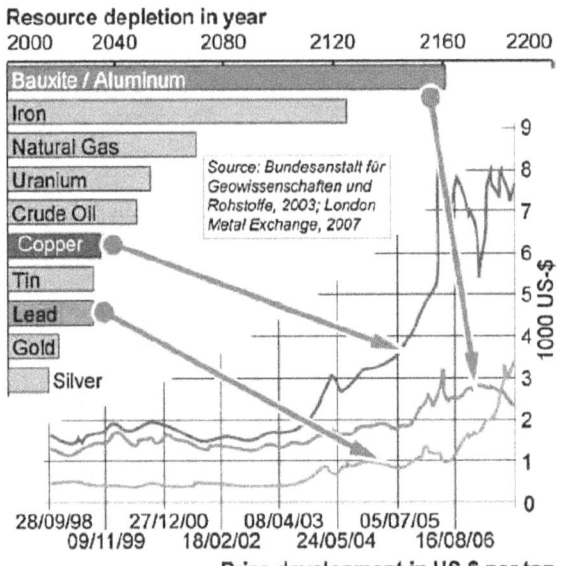

Figure 2. Natural resource development cost trends [Jovane 2008]

Other natural resources, such as precious metals and crude oil continue, depleting as the world economy expands [Jovane 2008]. Figure 2 shows that current growth in copper, silver, gold, lead, tin, and copper mining will deplete resources by 2040, and their prices will increase dramatically if the rates of their usages are not changed and recycling program are not put in place.

In addition to resource shortage, there are increased environmental concerns on toxic and hazardous wastes released to the environment, such as lead, mercury, and cadmium from electric and electronic products such as light bulbs, televisions, computers, and monitors. Regulations, such as Restriction of Hazardous Substances (RoHS) [RoHS 2002], Waste Electric and Electronic Equipment (WEEE) [WEEE 2002], and Registration, Evaluation, Authorization, and Restriction of Chemicals (REACH) [REACH 2006], have been developed by national governments to prevent environmental pollutants and health problem-causing substances from being imported and exported.

Recycle and reuse of products is an effective way to decrease the depletion of natural resources and the release of toxic materials to the environment. For example, recycling aluminum saves more than half of energy used in aluminum production from bauxite. In primary aluminum production, alumina refining takes about 40 % of the energy because the primary smelting requires most of the energy use in the aluminum manufacturing process [Green 2007]. Secondary smelting of recycled ingots takes about 6 % of the energy used in producing primary ingots. Recycling aluminum saves energy and the environment.

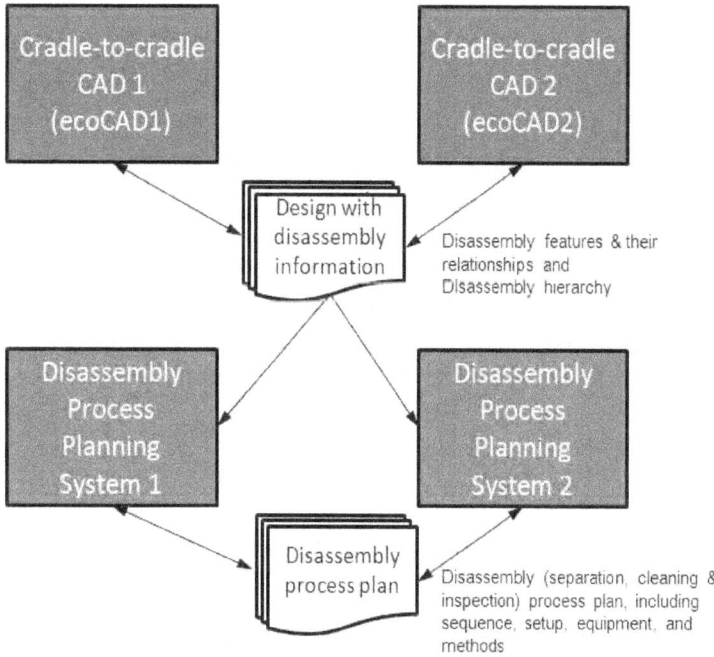

Figure 3. Disassembly information sharing

5

For reuse and recycle of a product at the end of its useful life, disassembly is the key operation to separate the product into reusable and recyclable parts. A disassembly process includes the separation of functional parts from the rest of the product for reuse, the cleaning of usable parts, and the inspection for the quality of the separated parts in reuse. To adequately plan for disassembly requires information from design, such as disassembly operation, sequence, disassembly features, feature relationships, and subassemblies. Disassembly process planning systems need also to exchange disassembly process plans, including task plan, equipment requirements, resource requirements, disassembly sequence, and cost estimates, among heterogeneous process planning systems. Designers require the cost of disassembly and available equipment to determine the disassemblability of a design. Designers also need to evaluate the ease of disassembly of a design. Figure 3 shows that design for disassembly data need to be exchanged among eco-Computer-Aided Design (ecoCAD) systems and disassembly process planning systems. Hence, an information model is necessary to share and exchange disassembly information among lifecycle applications, such as design and disassembly planning applications.

This paper describes an information model for disassembly defined using the Unified Modeling Language [Rumbaugh 1999]. Section 2 surveys the available literature on design for disassembly, disassembly process planning, and sustainable manufacturing. Section 3 describes all the classes and their relationships in the developed information model. Section 4 provides case studies to test the model. Section 5 gives possible future directions for information model development.

2. Current status of disassembly information modeling

Many available publications on why disassembly is important, disassembly representation in design, disassembly process plan representations, and feature modeling are reviewed in this section. Gaps towards an integrated disassembly information model are identified at the end of the section.

2.1 Motivations for disassembly

The need for protecting the environmental integrity and social well-being, as well as economic benefits, in manufacturing have been prominent as problems of global climate change, resource depletion, and unbalanced wealth distribution are worsening [Jovane 2008]. One proposed solution is to close the loop on material flows from manufacturers to users and back to manufacturers. The purpose is to reduce the extraction of resources from the earth and to reduce the energy use and pollution of producing stock materials [Nasr 2006]. The reuse and recycle of end-of-life products is the key to reduce landfill, energy in producing stock materials, and greenhouse gas emission [Sriram 2000]. Critical materials can be recovered from recycling the end-of-service-life products. In order to restrict solid and toxic wastes release to the environment, regulations on waste management on electronic products created by many countries around the world have been in place to curb the amount of toxic materials used in electronic products [Rifer 2009, Jofre 2005]. These regulations have already had positive impact on the environment and some economic impacts on the electronic industry due to the costs associated with conformance. The European Commission has placed regulations that 95% in weight of end-of-use vehicles will have to be recycled or recovered by 2015 [Kumar 2008]. Hence, disassembly for reuse, recycle, and remanufacturing becomes important. Design for disassembly becomes a needed research and development discipline to realize the concept of closed-loop material flows to protect the environment, reduce manufacturing costs, and decrease the rate of resources depletion [Bogue 2007, Jawahir 2008].

Also because of corporate image and customer demands, it is an industrial trend to look for means to manufacture products in a cleaner and more socially responsible way. Reuse, recycling, and remanufacturing promote sustainability through closed-loop material flows in a society. Disassembly is the key to achieve closed-loop material flows.

2.2 Design for disassembly

Many research results have been published in the areas of design for disassembly, disassembly process planning, and cost estimation. Disassemblability analysis of a product and the modeling of disassembly sequence have been developed in [Tang 2000-1]. An example of a possible disassembly path of components of an assembly is presented in [Lee 1996]. Disassemblability should be analyzed during product design so that the product may be timely and cost effectively disassembled at the end of service life. Algorithms have been developed to search for the optimal level of disassembling a product based on the cost and benefits of the extent that a product should be disassembled, i.e., the number of components into which a product should be disassembled [Desai 2003, Desai 2005]. Design analysis can be extended from disassemblability to remanufacturability, reusability, and recyclability [Ijomah 2007].

2.3 Modeling strategy for disassembly sequences and process planning

Disassembly methods can be divided into two types, namely non-destructive and destructive disassembly [Seliger 2007]. The non-destructive disassembly process includes loosening, unscrewing, desoldering, ungluing, and so on. The destructive disassembly processes are categorized into the groups - drilling, dividing (shearing and splitting), jet technologies (plasma arc cutting, water jet cutting, laser beam cutting), and inductive heating and thermal removal.

In a product disassembly system, choosing the representation of disassembly sequences is an important decision not only in creating a disassembly sequence planner but also in designing an intelligent controller for a disassembly process. Over the past decade, many modeling strategies have been proposed, i.e. AND/OR graph, disassembly Petri net (DPN), and Component-Fastener Graph. The AND/OR graph is the most popular and forms the basis for much of the later work. The nodes and the hyperarcs in these AND/OR graphs, respectively, correspond to subassemblies and disassembly tasks in which a more complex subassembly is separated into two subassemblies. Homem de Mello and Sanderson [Homem 1990] presented an AND/OR graph approach to model all the possible disassembly sequences of a system. The postulated condition is that it is possible to derive the assembly sequence by reversing the disassembly sequence obtained. For each possible assembly sequence generated, a certain assembly weight is assigned so that the optimal sequence could be obtained by comparing the weights assigned to the components. Lambert [Lambert 1997, Lambert 2003, Lambert 2005] used a so-called disassembly graph, which was based on the AND/OR graph and liaison analysis, to derive optimal disassembly sequences.

The disassembly Petri net approach is applied by Moore et al. [Moore 1998] for including complex AND/OR relationships in disassembly diagrams. Petri nets are frequently used in adaptive disassembly planners. The generation of an optimal disassembly sequence for devices with a probabilistic condition and adaptation based on additional observations has been a principal challenge for many years. Work on adaptive planners and some associated experimental results are in Zussman and Zhou [Zussman 2000]. An adaptive planner, based on product Petri nets and workstation Petri nets, which modifies the disassembly sequence according to the condition of the items in a batch, is presented by Tang et al. [Tang 2000-2]

Kuo et al. [Kuo 2000] propose a non-directed graph-based heuristic approach for the generation of the disassembly sequence for recycling. A product is modeled by a component-fastener graph. By identifying the "cut-vertices", the search splits the graph into subgraphs until a disassembly tree is formed. Based on the disassembly tree, disassembly sequences can then be generated. Li et al. [Li 2002, Li 2005] propose a Disassembly Constraint Graph (DCG), where all the possible disassembly operations that are needed for the maintenance of certain components or subassemblies can be deduced.

2.4 Information model for disassembly

Literature lacks an open, enabling information model of disassembly. We have reviewed an assembly information model for the purpose of creating a disassembly information model. Some standard-based approaches and frameworks are also reviewed.

ISO 10303-Part 44 [ISO10303 1994-1] provides some limited assembly design representations that capture assembly structure and kinematic joint information during the design process. The assembly model establishes a neutral representation of assemblies of products, which are composed of sets of components. In this model, complete products are called "assemblies," and the components of the lowest levels in the assemblies are called "parts." The model focuses on the hierarchy of the product, and on the position and orientation between parts.

The ISO working group (TC 184/SC4/WG12) (JNC proposal [ISO10303 2000]) has proposed several enhancements to the STandard for Exchange of Product data (STEP) assembly representation. In the Working Group (WG) 12 proposal, the detailed geometric information not only for hierarchical relationships but also for peer to peer relationships among component parts via an assembly feature is introduced. Geometric constraints among component parts at the detailed geometric element level are also enabled. The

WG12 proposal introduces more information on component association and includes detailed information about appropriate assembly features involved in component association.

An integrated National Institute of Standards and Technology - Core Product Model (NIST-CPM) [Fenves 2002] has been developed to unify and integrate product or assembly information. The NIST-CPM provides a base-level product model that is not tied to any vendor software. The model is open, non-proprietary, expandable, independent of any one product development process, and capable of capturing the engineering context that is most commonly shared in product development activities. The core model focuses on artifact representation, including function, form, behavior, and material. The model provides both physical and functional decompositions, and relationships among these concepts. The model is heavily influenced by the Entity-Relationship data model; accordingly, it consists of two sets of classes, called object and relationship, which is equivalent to the Unified Modeling Language (UML) class and association class, respectively.

The aim of the Open Assembly Model (OAM) [Sudarsan 2003] is to provide a standard representation and exchange protocol for assembly and system-level tolerance information. OAM is extensible; it currently provides tolerance representation and propagation, representation of kinematics, and engineering analysis at the system level. The assembly information model emphasizes the nature and information requirements for part features and assembly relationships. The model includes both assembly as a concept and assembly as a data structure. For the latter it uses the model data structures of STEP.

2.5 Feature information for disassembly

Few disassembly features and their representations have been studied [Kroll 1998]. Most feature representations are related to assembly, machining, and form features. They are applicable to disassembly.

De Fazio et al. [De Fazio 1993] propose a model called "Feature based Design for Assembly." In this model, the assembly features represented are the following: part features (location, volume, bounding box, material, and instances), part's shape features (location, orientation, dimensions, tolerances, surface texture, threaded surfaces, etc.), and assembled product's features (liaisons between parts and through which features, degrees of freedom, and distance between mating features). Shah [Shah 1992] has worked on the association between mating features of parts. His work deals with the determination of geometric constraints: degrees of freedom, compatibility between mating features, orientation and insertion limits. An assembly feature is defined as an association between two shape features of two different parts. Soldhi and Turner [Soldhi 1991] define an assembly feature as component features of mating components incorporated with tolerance information. They have a collection of elementary relations and matching form features, which are used for the generation of the assembly representation. Lee and Andrews [Lee 1985] focus on the spatial relationships imposed on the components in an assembly. Deneux [Deneux 1999] defines an assembly feature

9

as a generic solution referring to two groups of parts that need to be related by a relationship so as to solve a design problem. Holland and Bronsvoort [Holland 2000] define assembly features as "features with significance for assembly processes." The assembly features are subdivided into connection features and handling features. Chan and Tan [Chan 2003] define an assembly feature as the elementary connection feature containing mating relations between the components. Hamidullah et al. [Hamidullah 2006] define assembly features and their representation using the concept of assembly intents, which not only specifies the information of assembly and/or mating relations with the connecting form features but also associates the connecting form features with other assembly-specific information, for example, assembly operations, and assembly degrees of freedom.

The concept of connector has been applied to assembly planning by a few researchers. Akagi et al. [Akagi 1980] emphasize the significance of fasteners (i.e. connectors) in generating an assembly sequence, and categorize connectors into four groups based on the properties of movability (fixed or movable) and assemblability (disassembled or non-assembled). Furthermore, Gui and Mantyla [Gui 1994] propose a framework for assembly modeling in which the conceptual product building blocks are classified into two parts: the component and the connector. The above research works have focused primarily on assembly modeling with the connector concept. Tseng and Li [Tseng 1999] firstly decompose an assembly into a set of connector-based assembly elements, and then generate a connector-based assembly sequence without attempting to use any heuristic information provided by the connector. In their research, the concept of connector-based assembly element is defined, not to support reuse but to decrease the number of components involved in assembly sequencing. In Yin's paper [Yin 2003], two categories of connectors are identified from the perspective of their role in assembling, namely fixing and constraining. Generally, the only function of connectors for fixing is to combine initially disconnected components into a stable assembly. With respect to their fastening status, connectors for fixing can be further classified into two types: disassembled and non-assembled. For example, a screw belongs to the disassembled connector type since it can be taken apart easily with a screwdriver. However, disassembling a non-assembled connector like a rivet would be extremely difficult. Besides the connection function, connectors for constraining also provide other functions such as motion constraint or motion force transmission.

Zha and Du [Zha 2002] present a STEP-based method and system for concurrent integrated design and assembly planning. An integrated object model for mechanical systems and assemblies is first defined by a hierarchy of structure, geometry and feature. The structure is represented as a component-connector or joint multi-level graph with both hierarchical functional and assembly relations. These hierarchical relation models are then used for uniformly describing their relations both for assembly level and feature based single part level.

ISO 10303-Part 224 [ISO10303 2001] defines product data necessary for manufacturing a single piece or assembly of mechanical parts. It has machining features such as hole and pocket, and transition features such as round, fillet, and chamfer. ISO 10303-Part 111

[ISO10303 2007] specifies resource constructs for representing the complex shape elements, sometimes known as form features that are supported by the solid modeling capabilities of modern CAD systems. It has depression features such as hole and pocket, and edge blended features such as edge blend and chamfered edge.

2.6 High-level requirements for an information model

Based on the literature review, the following gaps are identified for modeling the information of integrated design for disassembly and disassembly process planning:

- A comprehensive modeling on simple, compound, and pattern disassembly features and separation between features from two component parts to be separated.
- Modeling the information of disassembly task sequence that is based on a disassembly sequence.
- Modeling the information of disassembly equipment and methods that are associated with the serial and parallel tasks and task decomposition in disassembly processes.
- Modeling the information of the cleaning and inspection process following the separation of assemblies of an end-of-service-life product.

3. Disassembly Information Model

This section describes all the classes and their relationships in the developed information model. The model has six major packages (modules). These packages support processes and operations of separation, cleaning, and inspection in a disassembly process. They are the Support Data package, the Feature package, the Tolerance package, the Workpiece package, the Equipment package, and the Workflow package.

3.1 Support Data

All the support data classes, including one subpackage PlacementPack, are in the Support Data Package in UML. The purpose of Support Data Package is to support classes in other packages in the disassembly information model. Default data types, such as character (char), boolean, integer, double, and float, in the UML are used throughout the model. Figure 4 shows the diagram of all the classes and the PlacementPack sub-package of Support Data.

Class String is used to represent a list of one or more characters. It has one attribute. Attribute chars is a list of **char**[1] of the type of Character, which is defined in the UML.

Class Identification is used to represent the identification of an object. It has one attribute. Attribute *theID*[2] is of type String.

---

[1] A type in bold italic font denotes a UML defined type.
[2] An attribute of a class in the disassembly information model is in italic font.

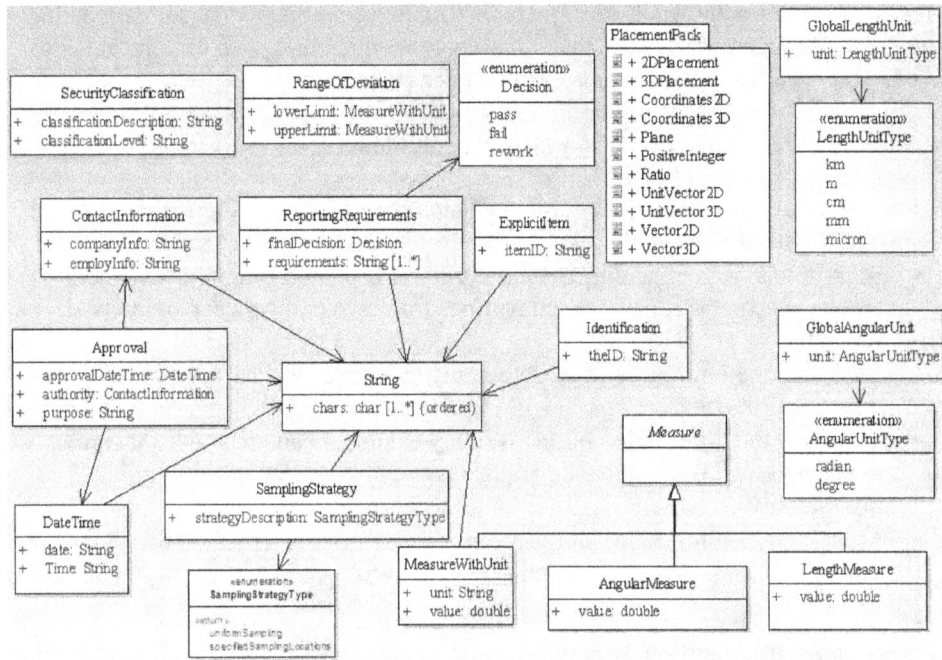

Figure 4. Class diagram of SupportDataPack

Class SamplingStrategy[3] is used to represent a description of sampling strategy in inspection using a specific probe and measuring machine. It has one attribute. Attribute *strategyDescription* is an enumeration type of SamplingStrategyType.

Enumeration type SamplingStrategyType is used to indicate the sampling strategy on measuring a part. This enumeration type includes *uniformSampling* and *specifiedSamplingLocations*.

Class DateTime is used to represent the date and time of an event. Both attributes *date* and *time* are of type String. Date and time should conform to ISO 8601 [ISO 8601].

Class ContactInformation is used to represent the information about a contact person. It has two attributes. Attribute *companyInfo* represents the information of the company of the contact person. Its type is String. Attribute *employInfo* represents the employee information of the contact person. Its type is also String.

Class Approval is used to represent the information about the approval of a disassembly process plan. It has three attributes. Attribute *approvalDateTime* represents the date and time that the process plan is approved, and its data type is DateTime. Attribute *authority* represents the person who approves the plan, and its data type is ContactInformation.

---

[3] The first letter of each word in a class name is capitalized.

Attribute *purpose* represents the description of the purpose of the plan, and its data type is String.

Enumeration type Decision is used to indicate the decision made on an inspected part. This enumeration type includes *pass*, *fail*, and *rework*.

Class ReportingRequirements is used to represent the requirement on reporting the results from the inspection of a part. It has two attributes. Attribute *finalDecision* represents the pass, fail, or rework of the inspected part, and its data type is Decision. Attribute *requirements* represents the description of reporting requirements, and its data type is String.

Class ExplicitItem is used to represent the geometric item's identification of a piece of geometry on a solid model that is identified as a feature. Attribute *itemID* represents the identification of a piece of geometry, such as a cylindrical surface representing a hole. The attribute's data type is String.

Class SecurityClassification is used to represent the security level of a process plan, such as a disassembly process plan or an inspection process plan. It has two attributes. Attribute *classificationDescription* represents a description of the security classification, and its data type is String. Attribute *classificationLevel* represents the level of classification of the process plan, and its data type is String.

Class Measure is an abstract data type. It has subtypes of LengthMeasure, MeasureWithUnit, and AngularMeasure.

Enumeration type LengthUnitType includes five commonly used length units, *km*, *m*, *cm*, *mm*, and *micron*.

Class GlobalLengthUnit is used to represent the length unit in a disassembly model. It has one attribute. Attribute *unit* represents the global unit of length measure. Its data type is LengthUnitType.

Class LengthMeasure is used to represent the measurement of a length. It has one attributes. Attribute *value* represents the value of the length measure, and its data type is double, which is a UML data type. The length unit is specified in the instance of GlobalLengthUnit.

Enumeration type AngularUnitType includes two commonly used angular units, *radian* and *degree*.

Class GlobalAngularUnit is used to represent the angular unit in the disassembly model. It has one attribute. Attribute *unit* represents the global unit of angular measure.

Class AngularMeasure is used to represent the measurement of an angle. It has one attribute. Attribute *value* represents the value of angular measure, and its data type is double, which is a UML data type. The angular unit is specified in the instance of GlobalAnglarUnit.

Class MeasureWithUnit is used to represent the measurement of a general measurand. It has two attributes. Attribute *unit* represents the unit of measure, and its data type is String. Attribute *value* represents the value of the measure, and its data type is **double**, which is a UML data type.

Class RangeOfDeviation is used to specify the lower limit (e.g., -0.2 mm) and the upper limit (e.g., +0.5 mm) of a global tolerance in a drawing or a 3D design model. The class has two attributes. Attribute *lowerLimit* represents the lower limit of the global tolerance, and its data type is MeasureWithUnit. Attribute *upperLimit* represents the upper limit of the global tolerance, and its data type is MeasureWithUnit.

The SupportData package includes a subpackage PlacementPack. Classes that are used to place a feature on a 3D workpiece are in the subpackage. Figure 5 shows a diagram of all the classes in this PlacementPack subpackage.

Class Ratio is used to represent the ratio of two quantities. It has one attribute. Attribute *x* represents the ratio, and its data type is **float**, which is a UML data type.

Class PositiveInteger is used to represent an integer that is greater than zero. It has one attribute. Attribute *x* represents the positive integer, and its data type is **int**, which is a UML data type. There is a constraint that *x* should be greater than zero.

Class Coordinates3D is used to represent coordinates of a point in the 3D space. It has three attributes. Attributes *x*, *y*, and *z* represent the 3 coordinates in the space, and their data types are **double**.

Class UnitVector3D is used to represent a unit vector in the 3D space. It has three attributes. Attributes *x*, *y*, and *z* represent the 3 components of the unit vector, and their data types are **double**. There is a constraint that the magnitude of the vector should be one.

Class Vector3D is used to represent a vector in the 3D space. It has two attributes. Attribute *origin* represents the starting point of the vector, and its data type is Coordinates3D. Attribute *end* represents the end point of the vector, and its data type is also Coordinates3D. Both the *origin* and *end* are not coincident.

Class Plane is used to represent a plane in the 3D space. It has two attributes. Attribute *location* represents the location point of the plane, and its data type is Coordinates3D. Attribute *normalVector* represents the orientation of the plane, and its data type is UnitVector3D.

Class 3DPlacement is used to represent the placement of a feature in the 3D space. It has three attributes. Attribute *location* represents the location of the feature, and its data type is Coordinates3D. Attribute *xDirection* represents the X direction of the 3D placement of

14

the feature, and its data type is Vector3D. Attribute yDirection represents the Y direction of the 3D placement of the feature, and its data type is also Vector3D. These two vectors must be mutually perpendicular. The Z direction is determined by the right-hand rule.

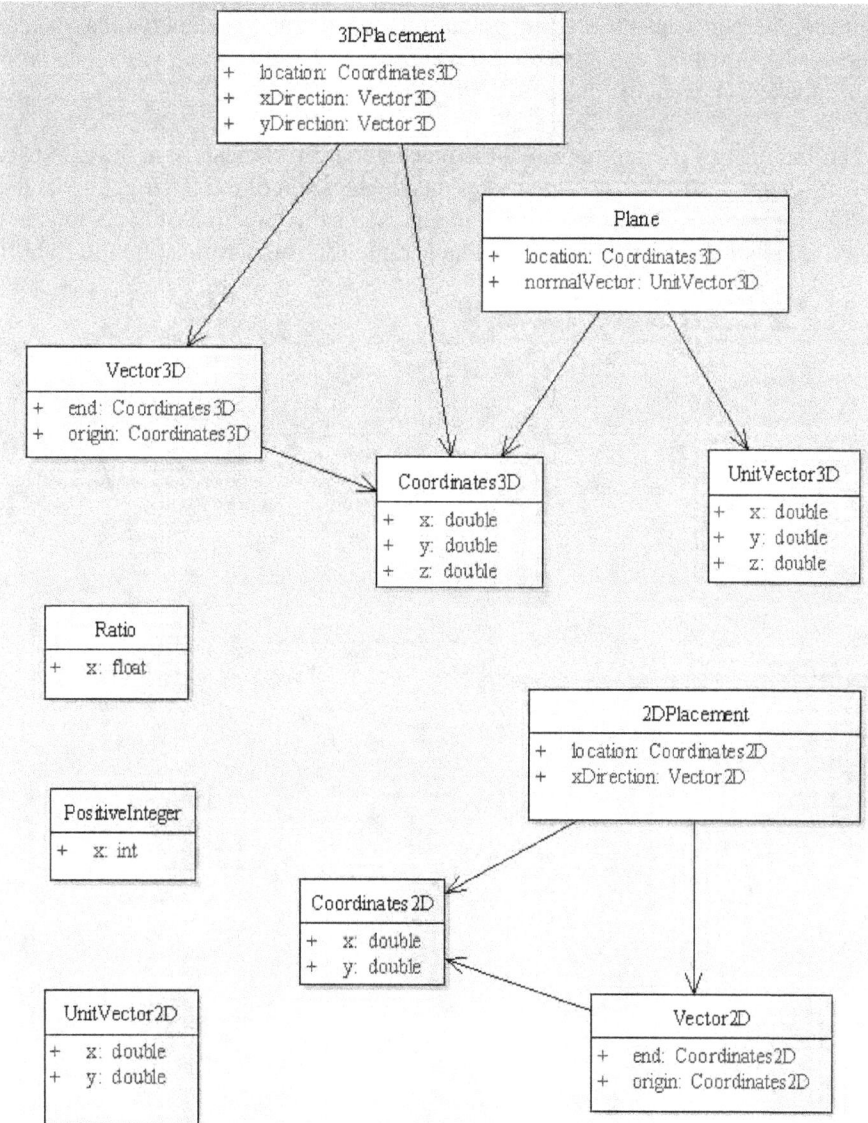

Figure 5. Class diagram of PlacementPack

Class Coordinates2D is used to represent coordinates of a point in the 2D space. It has two attributes. Attributes *x* and *y* represent the 2 coordinates in space, and their data types are ***double***.

Class UnitVector2D is used to represent a unit vector in the 2D space. It has two attributes. Attributes *x* and *y* represent the 2 components of the unit vector, and their data types are **double**. There is a constraint that the magnitude of the vector should be one.

Class Vector2D is used to represent a vector in the 2D space. It has two attributes. Attribute *origin* represents the starting point of the vector, and its data type is Coordinates2D. Attribute *end* represents the end point of the vector, and its data type is also Coordinates2D. Both the *origin* and *end* are not coincident.

Class 2DPlacement is used to represent the placement of a 2D feature in the 2D space. It has two attributes. Attribute *location* represents the location of the feature, and its data type is Coordinates2D. Attribute *xDirection* represents the direction of the X axis in the 2D space, and its data type is Vector2D. The Y direction is determined by the right-hand rule.

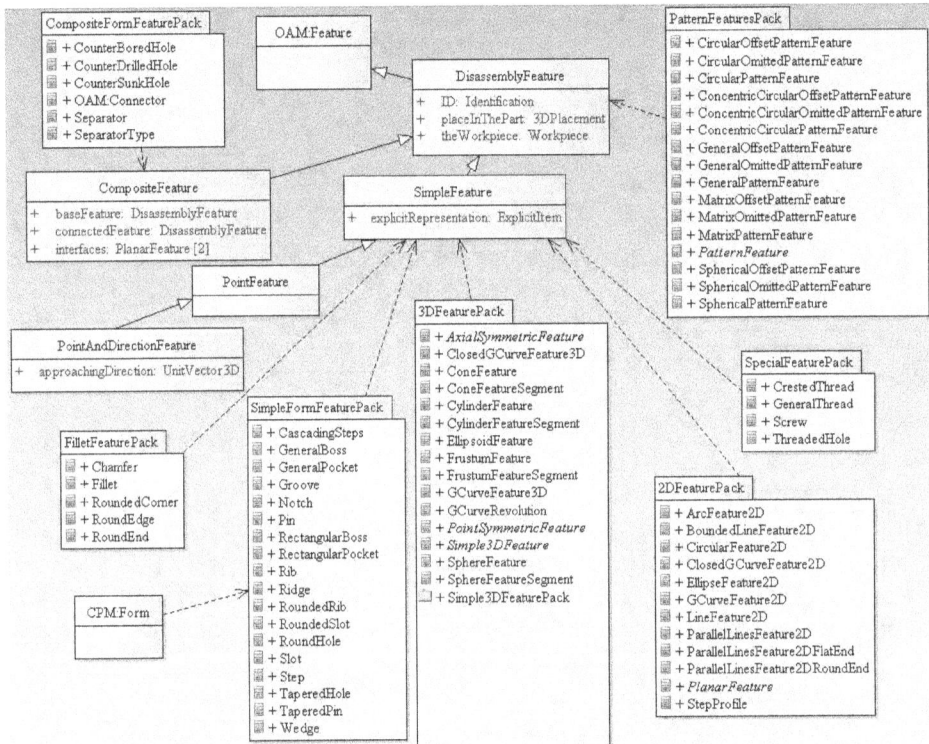

Figure 6. Class diagram of DisassemblyFeaturePack

3.2 Feature

A feature represents a specified portion of a part that is of interest in disassembly, including separation, cleaning, and inspection. From the complexity point of view, a simple feature is a baseline feature. From the geometric point of view, a simple feature can be a point, 2D, or 3D feature. Composite and pattern features are defined based on simple feature. Figure 6 shows the diagram of the DisassemblyFeaturePack. The package contains all the classes that are used to describe features relevant to disassembly. Specific feature representations for disassembly applications are in subpackages.

Class CPM:Feature is used to represent a feature defined in the Core Product Model. Disassembly feature is a subtype of the CPM:Feature class.

Class CPM:Form is used to represent the form of a feature. It is defined in the Core Product Model.

Class DisassemblyFeature is used to represent a feature that is involved in one or more operations of separation, cleaning, and inspection. It has three attributes. Attribute *ID* represents the feature identification, and its data type is Identification. Attribute *placeInThePart* represents the placement of the feature in the part model, and its data type is 3DPlacement. Attribute *theWorkpiece* represents the workpiece on a machine tool, and its data type is Workpiece (to be defined in the Workpiece package).
Class SimpleFeature is a subtype of DisassemblyFeature and used to represent a simple, generic feature that serves as the building block for more complex features. It has one attribute. Attribute *explicitRepresentation* represents the geometric and topological item that defines the geometry of the simple feature, and its data type is ExplicitItem. ExplicitItem is defined in ISO 10303-Part 203 [ISO10303 1994-2].

Class PointFeature is a subtype of SimpleFeature used to represent a point that is used in a disassembly process, such as inspection of a feature in the workpiece.

Class PointAndDirectionFeature is a subtype of PointFeature and used to represent a point with a direction that is used in an inspection process. It has one attribute. Attribute *approachingDirection* represents the direction of approaching the point feature, and its data type is UnitVector3D.

Subpackages in the Disassembly Feature Package include 2DFeaturePack, 3DFeaturePack, SimpleFormFeaturePack, FilletFeaturePack, SpecialFeaturePack, CompositeFormFeaturePack, and PatternFeaturePack. They are described in the following subsections.

3.2.1 Simple 2D Features

Simple 2D features are planar features and included in the 2DFeaturePack subpackage. Figure 7 shows the diagram of all the planar feature classes in 2DFeaturePack.

Class PlanarFeature is an abstract class and used to represent a 2D feature on a plane. The class is a subtype of DisassemblyFeature and has one attribute. Attribute *2DFeaturePlacement* represents the plane where the feature is placed on a workpiece, and its data type is Plane.

Class LineFeature2D is a subtype of PlanarFeature and used to represent a line on a flat surface. It has two attributes. Attribute *direction* represents the direction of the line

17

feature, and its data type is UnitVector2D. Attribute *pivot* represents a point on the line feature, and its data type is Coordinates2D.

Class BoundedLineFeature2D is a subtype of PlanarFeature and used to represent a line segment with two end points on a flat surface. It has one attribute. Attribute *endPoints* represents the two end points of a bounded line feature, and its data type is an array of Coordinates2D.

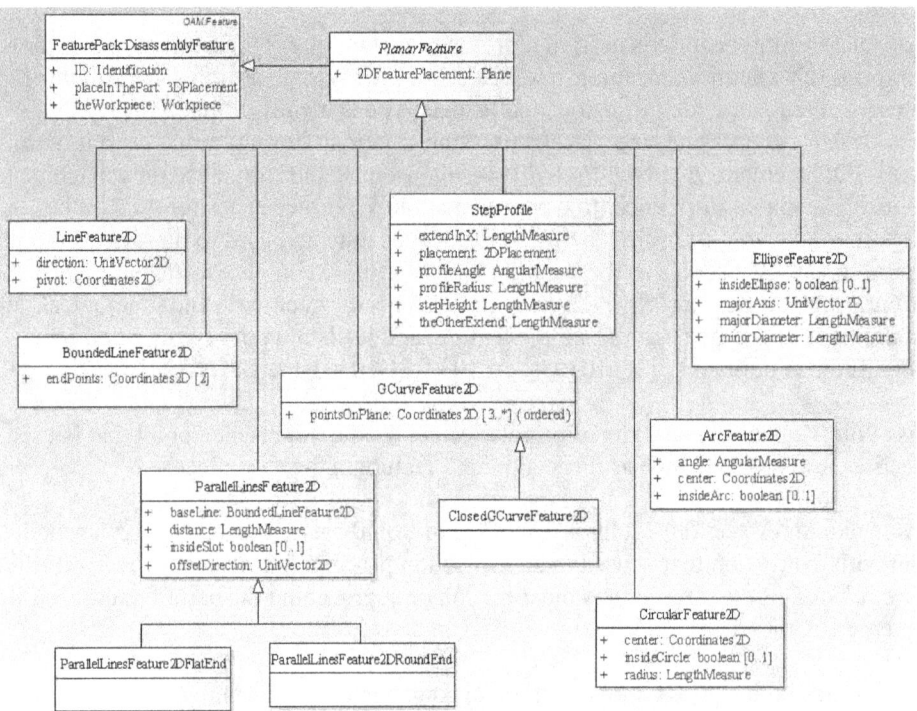

Figure 7. Class diagram of simple 2D features in 2DFeaturePack
NOTE: class name in italic font in a UML class diagram means the class is abstract.

Class ParallelLinesFeature2D is an abstract class and a subtype of PlanarFeature. ParallelLinesFeature2D is used to represent a pair of parallel lines on a flat surface. It has four attributes. Attribute *baseLine* represents the line feature that is referenced by the other line, and its data type is BoundedLineFeature2D. Attribute *distance* represents the distance between the two parallel lines, and its data type is LengthMeasure. Attribute *insideSlot* represents the indication whether the feature is inside of a slot or not. If yes, it is a depression, otherwise, it is a protrusion. Its data type is ***boolean***, which is a basic data type in UML. Attribute *offsetDirection* represents the direction of the other line relative to the base line, and its data type is UnitVector2D. The unit vector has to be perpendicular to the base line feature.

Class ParallelLinesFeature2DFlatEnd is a subtype of ParallelLinesFeature2D and used to represent the feature that the two ends of the parallel lines are connected to each other by two parallel line segments. The feature is rectangle.

Class ParallelLinesFeature2DRoundEnd is also a subtype of ParallelLinesFeature2D and used to represent a pair of parallel lines with two round ends. Each round end is an arc of the diameter equal to the distance of the two parallel lines. The arcs must make the feature convex, not concave.

Class GCurveFeature2D is a subtype of PlanarFeature and used to represent a general curve feature on a flat surface. It has one attribute. Attribute *pointsOnPlane* represents the control points that are used to generate the curve, and its data type is a list of Coordintes2D with minimum of three elements in the list. A two dimensional non-uniform B-spline curve is assumed.

Class ClosedGCurveFeature2D is a subtype of GCurveFeature2D and used to represent a closed-end general curve feature on a flat surface. The order of continuity at the closing end is the order of the curve (n) minus 1 ($= n - 1$).

Class StepProfile is used to represent a feature that is a 2D step. It has six attributes. Attribute *extendInX* represents the extent of the first portion of the step, and its data type is LengthMeasure. Attribute *placement* represents the placement of the step feature on a flat surface, and its data type is 2DPlacement. Attribute *profileAngle* represents the angle of the step since the angle is not necessarily vertical, and its data type is AngularMeasure. Attribute *profileRadius* represents the radius of the rounded corner, and its data type is LengthMeasure. Note that if the radius is zero, it means that there is no rounded corner. Attribute *theOtherExtend* represents the extent of the third portion of the step, and its data type is LengthMeasure. Attribute *stepHeight* represents the height of the step, and its data type is LengthMeasure. An illustration of the step feature and its attributes can be found in ISO 10303-Part 224 [ISO10303 2001].

Class CircularFeature2D is used to represent a feature that is a circle. It has three attributes. Attribute *center* represents the center of the circle on the plane where the circle is placed. Its data type is Coordinates2D. Attribute *insideCircle* represents that the material is inside, i.e., a solid circular feature. Its data type is *boolean*. Attribute *radius* represents the radius of the circle, and its data type is LengthMeasure.

Class ArcFeature2D is used to represent a feature that is an arc. It has three attributes. Attribute *angle* represents the angle of the arc, and its data type is AngularMeasure. Attribute *center* represents the center of the arc on the plane where the arc is placed. Attribute *insideArc* represents that the material is inside the arc, and its data type is *boolean*.

Class EllipseFeature2D is used to represent a feature that is an ellipse. It has five attributes. Attribute *insideEllipse* represents the material side of the feature, and its data type is *boolean*. Attribute *majorAxis* represents the major axis of the ellipse, and its data type is UnitVector2D. The minor axis is perpendicular to the major axis. Attribute *majorDiameter* represents the major diameter of the ellipse, and its data type is LengthMeasure. Attribute *minorDiameter* represents the minor diameter of the ellipse, and its data type is LengthMeasure.

3.2.2 Simple 3D Features

All the simple 3D feature classes are in the subpackage of 3DFeaturePack. They are three dimensional. They are developed based on many simple 2D features, described in the previous section. Simple 3D features are fundamental building blocks for defining form, pattern, and special disassembly features. Figure 8 shows the diagram of 3DFeaturePack.

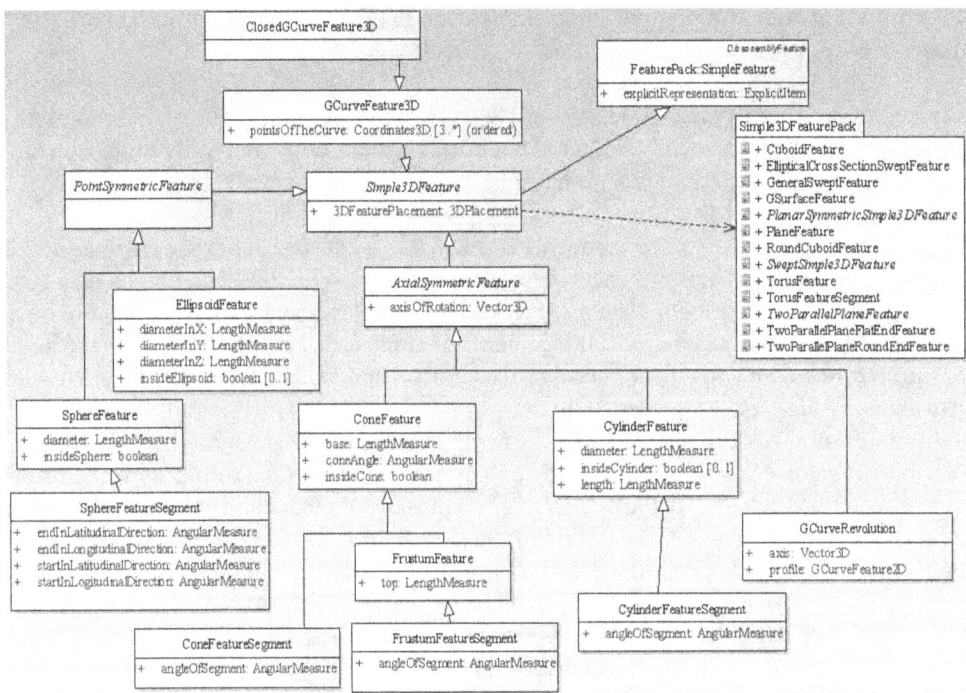

Figure 8. Class diagram of 3DFeaturePack

Class *Simple3DFeature* is abstract and used to represent a simple feature placed in a 3D space, e.g., a work space in a disassembly workstation or on an inspection machine. It has one attribute. Attribute *3DFeaturePlacement* represents the location of the feature and its orientation in the 3D space, and its data type is 3DPlacement. Every Simple3DFeature is a surface, not a solid.

Class PointSymmetricFeature is abstract and used to represent a feature symmetric about a point placed in a 3D space. The class is a subtype of *Simple3DFeature*. The point is the center of the feature and located at the origin of the coordinate system located by the *3DFeaturePlacement*.

Class SphereFeature is used to represent a feature that is a sphere placed in a 3D space and is a subtype of PointSymmetricFeature. SphereFeature has two attributes. Attribute *diameter* represents the size of the sphere, and its data type is LengthMeasure. Attribute *insideSphere* represents that the material is inside the sphere, and its data type is optional **boolean**. True represents material inside, i.e., a solid sphere.

Class SphereFeatureSegment is used to represent a feature that is a spherical segment placed in a 3D space. It is a subtype of SphereFeature. The latitude direction starts from the positive X axis on the X-Y plane rotating towards the positive Y axis. The longitude direction starts from the positive Z axis on the Z-X plane rotating towards the positive X axis. The class has four attributes. Attribute *endInLatitudeDirection* represents the angle of the end of the spherical segment in the latitude direction of a spherical coordinates system, and its data type is AngularMeasure. Attribute *endInLongitudeDirection* represents the angle of the end of the spherical segment in the longitude direction of a spherical coordinates system, and its data type is AngularMeasure. Attribute *startInLatitudeDirection* represents the angle of the start of the spherical segment in the latitude direction of a spherical coordinates system, and its data type is AngularMeasure. Attribute *startInLongitudeDirection* represents the angle of the start of the spherical segment in the longitude direction of a spherical coordinates system, and its data type is AngularMeasure.

Class EllipsoidFeature is used to represent a feature that is an ellipsoid placed in a 3D space and is a subtype of PointSymmetricFeature. EllipsoidFeature has four attributes. Attribute *diameterInX* represents the diameter of the ellipsoid in the X direction, and its data type is LengthMeasure. Attribute *diameterInY* represents the diameter of the ellipsoid in the Y direction, and its data type is LengthMeasure. Attribute *diameterInZ* represents the diameter of the ellipsoid in the Z direction, and its data type is LengthMeasure. Attribute *insideEllipsoid* represents that the material is inside of the ellipsoid, and its data type is optional **boolean**. True represents material inside, i.e., a solid ellipsoid.

Class *AxialSymmetricFeature* is an abstract class and used to represent a feature symmetric about an axis placed in a 3D space. The class is a subtype of *Simple3DFeature*. It has one attribute. Attribute *axisOfRotation* represents the axis of symmetry, and its data type is Vector3D.

Class ConeFeature is used to represent a feature that is a cone placed in a 3D space. It has three attributes. The class is a subtype of *AxialSymmetricFeature*. Attribute *base* represents the diameter of the cone base, and its data type is LengthMeasure. Attribute *coneAngle* represents the angle of the cone, and its data type is AngularMeasure. The *coneAngle* must be positive and less than 180 degrees. Attribute *insideCone* represents the material side of the cone, and its data type is **boolean**. If the value is false, it is a protrusion, otherwise, it is a depression. The cone has its base on the XY plane and its tip on the positive Z axis.

Class ConeFeatureSegment is used to represent a feature that is a cone segment placed in a 3D space. The class is a subtype of ConeFeature. ConeFeatureSegment has one attribute. Attribute *angleOfSegment* represents the angle of the cone segment, and its data type is AngularMeasure. The *angleOfSegment* starts from the positive X-axis on the X-Y plane. The ConeFeatureSegment is the points on the cone lying inside the solid of revolution made by rotating the positive ZX half plane about the Z-axis through the *angleOfSegment*.

Class FrustumFeature is used to represent a feature that is a frustum in a 3D space. It is a subtype of ConeFeature. FrustumFeature has one attribute. Attribute *top* represents the diameter of the top of the frustum, and its data type is LengthMeasure.

Class FrustumFeatureSegment is used to represent a feature that is a segment of a frustum in a 3D space. It is a subtype of FrustumFeature. FrustumFeatureSegment has one attribute. Attribute *angleOfSegment* represents the angular size of the frustum segment, and its data type is AngularMeasure. The *angleOfSegment* starts from the positive X-axis on the X-Y plane. The FrustumFeatureSegment is the points on the frustum lying inside the solid of revolution made by the positive ZX half plane about the Z-axis through the *angleOfSegment*.

Class CylinderFeature is used to represent a feature that is a cylinder in a 3D space. The class is a subtype of *AxialSymmetricFeature*. It has three attributes. Attribute *diameter* represents the diameter of the cylinder, and its data type is LengthMeasure. Attribute *insideCylinder* represents the material side of the cylinder, and its data type is optional **boolean**. True represents material inside. Attribute *length* represents the length of the cylinder, and its data type is also LengthMeasure.

Class CylinderFeatureSegment is used to represent a feature that is a cylindrical segment in a 3D space. The class is a subtype of CylinderFeature. It has one attribute. Attribute *angleOfSegment* represents the angular size of the cylinder segment, and its data type is AngularMeasure. The angle is measured from the positive X axis on the XY plane.

Class GCurveFeature3D is used to represent a feature that is a general curve in a 3D space. The class is a subtype of *Simple3DFeature*. It has one attribute. Attribute *pointsOfTheCurve* represents the control points that are used to generate the curve, and its data type is a list of Coordinates3D with minimum of three elements in the list. The non-uniform rational B-spline curve is assumed.

Class ClosedGCurveFeature3D is a subtype of GCurveFeature3D and used to represent a closed-end general curve feature in a 3D space. The order of continuity at the closing end is the order of the curve (n) minus 1 (= n − 1).

Class GCurveRevolution is used to represent a surface that is created by revolving a 2D curve about an axis. The class is a subtype of *AxialSymmetricFeature*. It has two attributes. Attribute *axis* represents the axis of revolution that is used to generate the surface, and its data type is Vector3D. Attribute *profile* represents the 2D curve, and its data type is GCurveFeature2D.

Figure 9 shows the diagram of the Simple3DFeaturePack subpackage that includes the following simple 3D feature classes.

Class PlaneFeature is a subtype of *Simple3DFeature* and used to represent a feature that is a rectangle placed in a 3D space. It has two attributes. Attribute *length* represents the length of the plane, and its data type is LengthMeasure. Attribute *width* represents the width of the plane, and its data type is also LengthMeasure. The rectangle lies on the XY plane of the coordinate system located by the *3DFeaturePlacement*. The center of the rectangle is at the origin, length is parallel to the X axis, and width is parallel to the Y axis.

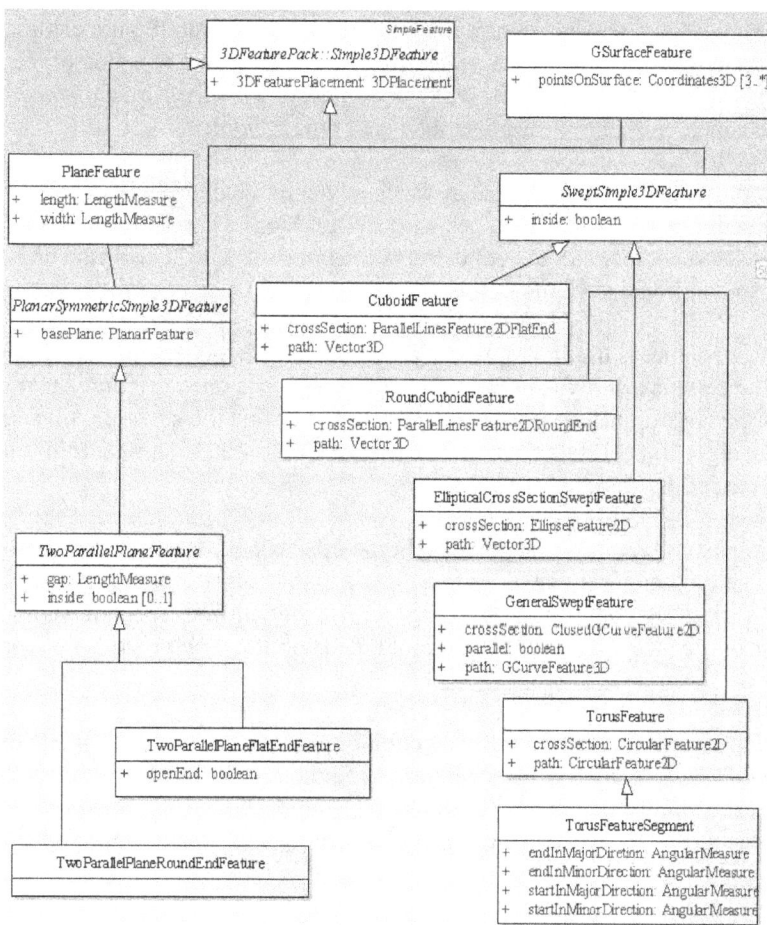

Figure 9. Class diagram of Simple3DFeaturePack

Class PlanarSymmetricSimple3DFeature is abstract, a subtype of *Simple3DFeature*, and used to represent a feature that is symmetric about a plane. It has one attribute. Attribute *basePlane* represents the plane about which the feature is symmetric, and its data type is PlaneFeature.

Class TwoParallelPlaneFeature is abstract, a subtype of PlanarSymmetricSimple3DFeature, and used to represent a pair of parallel planes in a 3D space. It has two attributes. Attribute *gap* represents the distance between the two parallel planes, and its data type is LengthMeasure. (Note: the two parallel planes are symmetric about the base plane.) The sizes of the two parallel planes must be the same and are determined by the size of the base plane. Attribute *inside* represents that the material side of the swept feature, and its data type is optional ***boolean***. True represents that the material is inside.

Class TwoParallelPlaneFlatEndFeature is a subtype of TwoParallelPlaneFeature and is used to represent a pair of parallel planes with a closed flat end at one side of the two parallel planes and another closed flat end at the opposite side. Attribute *openEnd* represents the open end of the feature, and its data type is ***boolean***.

Class TwoParallelPlaneRoundEndFeature is a subtype of TwoParallelPlaneFeature and used to represent a pair of parallel planes with a closed round end at one side of the two parallel planes and another closed round end at the opposite side. The round ends have to be convex. No additional attribute in this class.

Class GSurfaceFeature is used to represent a feature that is a general surface placed in a 3D space. The class is a subtype of *Simple3DFeature*. It has one attribute. Attribute *pointsOnSurface* represents the points that are used to define the surface with a specified mathematical algorithm, and its data type is a set of Coordinates3D, with minimum 3 elements in the set.

Class *SweptSimple3DFeature* is abstract, subtype of *Simple3DFeature*, and used to represent a feature that is generated by sweeping a planar surface as the cross-section in space using a specified path. It has one attribute. Attribute *inside* represents that the material side of the swept feature, and its data type is optional ***boolean***. True represents material inside.

Class CuboidFeature is used to represent a cube in a 3D space. The class is a subtype of *SweptSimple3DFeature*. It has two attributes. Attribute *crossSection* represents the cross section of a cube, and its data type is ParallelLineFeature2DFlatEnd. The *path* is the linear path that the cross section is swept, and its data type is Vector3D.

Class RoundCuboidFeature is used to represent a rounded cube in a 3D space. The class is a subtype of *SweptSimple3DFeature*. RoundCuboidFeature has two attributes. Attribute *crossSection* represents the cross section of a cube, and its data type is ParallelLinesFeature2DRoundEnd. The *path* is the linear path that the cross section is swept, and its data type is Vector3D.

Class EllipticalCrossSectionSweptFeature is used to represent a surface feature that is generated by sweeping an ellipse in a 3D space with a linear path. The class is a subtype of *SweptSimple3DFeature*. EllipticalCrossSectionSweptFeature has two attributes. Attribute *crossSection* represents the cross section of the feature, and its data type is EllipseFeature2D. The *path* is the linear path that the cross section is swept, and its data type is Vector3D.

Class GeneralSweptFeature is used to represent a swept feature that is created by sweeping a cross-section along a path. The class is a subtype of *SweptSimple3DFeature*. GeneralSweptFeature has three attributes. Attribute *crossSection* represents the cross-section that is used to generate the volume, and its data type is ClosedGCurveFeature2D. The cross-section feature is placed at the beginning of the sweeping path. Attribute *path* represents the sweeping path, and its data type is GCurveFeature3D. Attribute *parallel* represents the sweeping cross-section is parallel to the cross-section feature, and its data type is **boolean**. If *parallel* is false, the cross-sectional plan is perpendicular to the sweeping path.

Class TorusFeature is used to represent a feature that is a torus placed in a 3D space. The class is a subtype of *SweptSimple3DFeature*. It has two attributes. Attribute *crossSection* represents the cross section of the feature, and its data type is CircularFeature2D. The *path* is the circular path that the cross section is swept, and its data type is CircularFeature2D.

Class TorusFeatureSegment is used to represent a feature that is a torus segment placed in a 3D space. It is a subtype of TorusFeature. The feature is placed so that the major direction starts from the positive X axis on the X-Y plane rotating towards the positive Y axis. The minor direction starts from the X-Y plane rotating towards the positive Z axis. It has four attributes. Attribute *endInMajorDirection* represents the angle of the end of the torus segment in the major diameter direction, and its data type is AngularMeasure. Attribute *endInMinorDirection* represents the angle of the end of the torus segment in the minor diameter direction, and its data type is AngularMeasure. Attribute *startInMajorDirection* represents the angle of the start of the torus segment in the major diameter direction, and its data type is AngularMeasure. Attribute *startInMinorDirection* represents the angle of the start of the torus segment in the minor diameter direction, and its data type is AngularMeasure.

3.2.3 Simple Form Features

Based on simple 3D features, specific form features are defined for applications in design for disassembly and disassembly process planning at the end of a product's useful life. They are all subtypes of SimpleFeature, except CascadingSteps. Figure 10 shows the diagram of all the simple form feature classes in the SimpleFormFeaturePack.

Class RoundHole is used to represent a hole in a workpiece. It has one attribute. Attribute *theHole* represents the hole, and its data type is CylinderFeature.

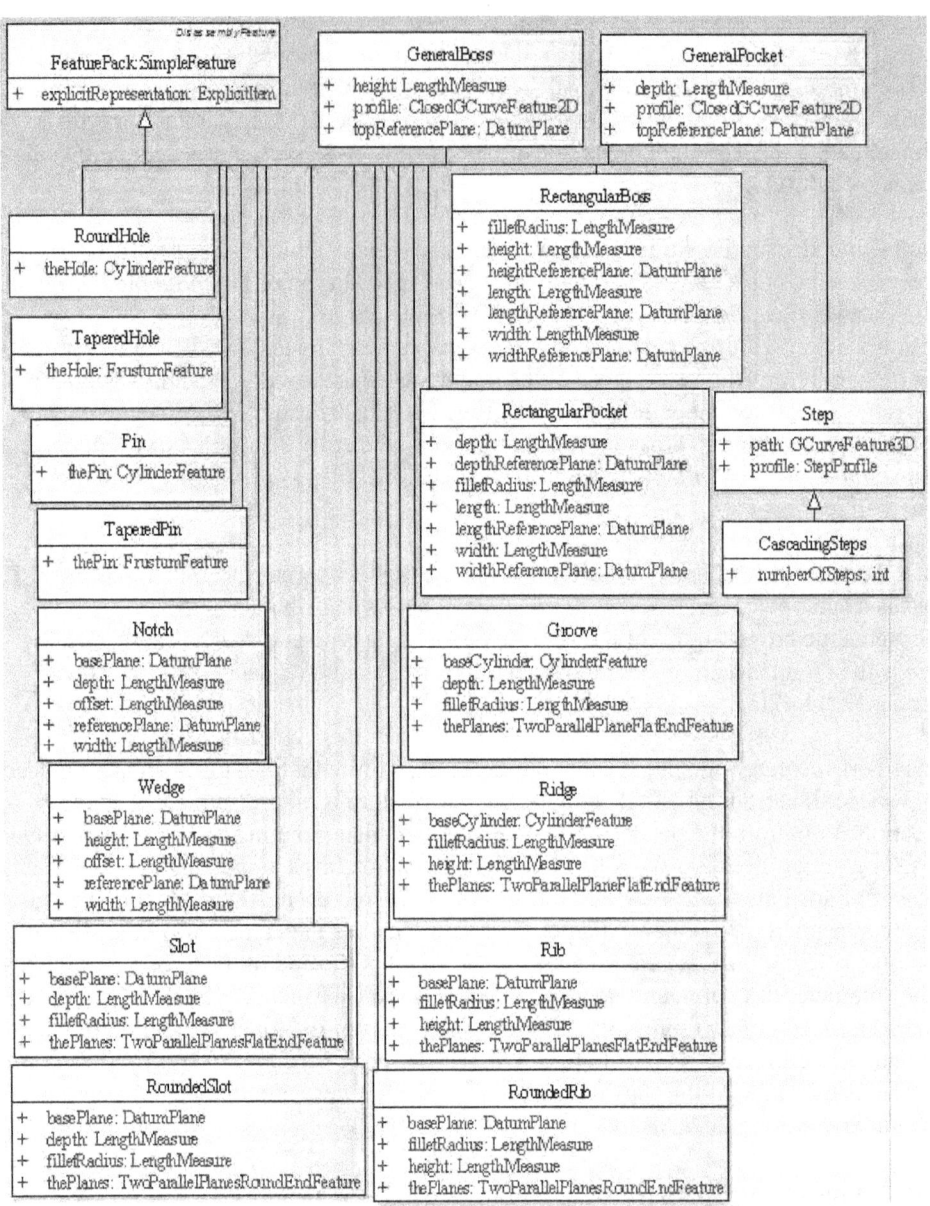

Figure 10. Class diagram of SimpleFormFeaturePack

Class TaperedHole is used to represent a tapered hole in a workpiece. It has one attribute. Attribute *theHole* represents the tapered hole, and its data type is FrustumFeature.

Class Pin is used to represent a pin (a protrusion) in a workpiece. It has one attribute. Attribute *thePin* represents the pin, and its data type is CylinderFeature.

Class TaperedPin is used to represent a tapered pin in a workpiece. It has one attribute. Attribute *thePin* represents the tapered pin, and its data type is FrustumFeature.

Class Slot is used to represent a slot, a depression. It has four attributes. Attribute *basePlane* represents the base plane from which the slot is depressed, and its data type is DatumPlane. (DatumPlane is defined in Section 3.3.1.) Attribute *depth* represents the depth of the slot, measured from the base plane, and its data type is LengthMeasure. Attribute *filletRadius* represents the fillet radius of the edges that fillets can be applied to edges of the slots, and its data type is LengthMeasure. Attribute *thePlanes* represents two parallel planes that form the sides of the slot, and its data type is TwoParallelPlanesFlatEndFeature.

Class RoundedSlot is used to represent a slot of rounded ends. It has four attributes. Attribute *basePlane* represents the base plane from which the slot is depressed, and its data type is DatumPlane. Attribute *depth* represents the depth of the slot, measured from the base plane, and its data type is LengthMeasure. Attribute *filletRadius* represents the fillet radius of the edges that fillets can be applied to edges of the slot, and its data type is LengthMeasure. Attribute *thePlanes* represents two parallel planes with round ends that form the boundaries of the slot, and its data type is TwoParallelPlanesRoundEndFeature.

Class Rib is used to represent a rib, a protrusion. It has four attributes. Attribute *basePlane* represents the base plane from which the rib protrudes, and its data type is DatumPlane. Attribute *filletRadius* represents the fillet radius of the edges that fillets can be applied to edges of the rib, and its data type is LengthMeasure. Attribute *depth* represents the height of the rib, and its data type is LengthMeasure. Attribute *thePlanes* represents two parallel planes that form the sides of the rib, and its data type is TwoParallelPlanesFlatEndFeature.

Class RoundedRib is used to represent a rib with rounded ends. It has four attributes. Attribute *basePlane* represents the base plane from which the rib protrudes, and its data type is DatumPlane. Attribute *filletRadius* represents the fillet radius of the edges that fillets can be applied to edges of the rib, and its data type is LengthMeasure. Attribute *height* represents the height of the rib from the base plane, and its data type is LengthMeasure. Attribute *thePlanes* represents two parallel planes that form the boundaries of the rib with rounded ends, and its data type is TwoParallelPlanesRoundEndFeature.

Class Notch is used to represent a notch, a triangular slot. It has five attributes. Attribute *basePlane* represents the base plane from which the notch is depressed, and its data type is DatumPlane. Attribute *depth* represents the depth of the notch, and its data type is LengthMeasure. Attribute *offset* represents the offset distance from the reference plane to the center of the notch, and its data type is LengthMeasure. Attribute *referencePlane* represents a reference plane to define the center of the notch, and its data type is DatumPlane. Attribute *width* represents the width of the notch, and its data type is LengthMeasure.

Class Wedge is used to represent a wedge, with a triangular profile. It has five attributes. Attribute *basePlane* represents the base plane from which the wedge protrudes, and its data type is DatumPlane. Attribute *height* represents the height of the wedge, and its data type is LengthMeasure. Attribute *offset* represents the offset distance from the reference plane to the center of the wedge, and its data type is LengthMeasure. Attribute *referencePlane* represents a reference plane that is used to define the center of the wedge, and its data type is DatumPlane. Attribute *width* represents the width of the wedge, and its data type is LengthMeasure.

Class Ridge is used to represent a ridge (a protrusion) on an axially symmetric workpiece. It has four attributes. Attribute *baseCylinder* represents the base cylinder from which the ridge protrudes, and its data type is CylinderFeature. Attribute *filletRadius* represents the fillet radius of the edges that fillets can be applied on the ridge, and its data type is LengthMeasure. Attribute *height* represents the height of the ridge, and its data type is LengthMeasure. Attribute *thePlanes* represents two parallel planes that form the sides of the ridge, and its data type is TwoParallelPlanesFlatEndFeature.

Class Groove is used to represent a groove (a depression) in an axially symmetric workpiece. It has four attributes. Attribute *baseCylinder* represents the base cylinder from which the groove is depressed, and its data type is CylinderFeature. Attribute *filletRadius* represents the fillet radius of the edges of the groove, and its data type is LengthMeasure. Attribute *depth* represents the depth of the groove, and its data type is LengthMeasure. Attribute *thePlanes* represents two parallel planes that form the sides of the groove, and its data type is TwoParallelPlanesFlatEndFeature.

Class RectangularBoss is used to represent a boss (a protrusion) in a workpiece. It has seven attributes. Attribute *filletRadius* represents the fillet radius of the edges of the boss, and its data type is LengthMeasure. Attribute *height* represents the height of the boss, and its data type is LengthMeasure. Attribute *heightReferencePlane* represents the base plane that the boss is protruded, and its data type is DatumPlane. Attribute *length* represents the length of the rectangular boss, and its data type is LengthMeasure. Attribute *lengthReferencePlane* represents the datum plane that the length of the boss is referenced, and its data type is DatumPlane. Attribute *width* represents the width of the rectangular boss, and its data type is LengthMeasure. Attribute *widthReferencePlane* represents the datum plane that the width of the boss is referenced, and its data type is DatumPlane.

Class RectangularPocket is used to represent a pocket (a depression) in a workpiece. It has seven attributes. Attribute *filletRadius* represents the fillet radius of the edges of the boss, and its data type is LengthMeasure. Attribute *depth* represents the depth of the pocket, and its data type is LengthMeasure. Attribute *depthReferencePlane* represents the base plane from which the pocket is depressed, and its data type is DatumPlane. Attribute *length* represents the length of the rectangular pocket, and its data type is LengthMeasure. Attribute *lengthReferencePlane* represents the datum plane that the length of the pocket is referenced, and its data type is DatumPlane. Attribute *width* represents the width of the rectangular pocket, and its data type is LengthMeasure. Attribute *widthReferencePlane* represents the datum plane that the width of the pocket is referenced, and its data type is DatumPlane.

Class GeneralBoss is used to represent a general boss in a workpiece. It has three attributes. Attribute *height* represents the height of a general boss, and its data type is LengthMeasure. Attribute *topReferencePlane* represents the base plane from which the boss protrudes, and its data type is DatumPlane. Attribute *profile* represents the profile of the general boss, and its data type is ClosedGCurveFeature2D.

Class GeneralPocket is used to represent a general pocket in a workpiece. It has three attributes. Attribute *depth* represents the depth of a general pocket, and its data type is LengthMeasure. Attribute *topReferencePlane* represents the base plane from which the pocket is depressed, and its data type is DatumPlane. Attribute *profile* represents the profile of the general pocket, and its data type is ClosedGCurveFeature2D.

Class Step is used to represent a feature that is a step in a workpiece. It has two attributes. Attribute *profile* represents the profile of the step, and its data type is StepProfile. Attribute *path* represents the path of the sweeping of the step profile that is used to generate a step feature, and its data type is GCurveFeature3D.

Class CascadingSteps is used to represent a feature that is a cascade of steps in a workpiece. It has one attribute. Attribute *numberOfSteps* represents the number of a sequence of cascading steps in the feature, and its data type is *int* (a UML type of integer).

3.2.4 Fillet Features

Fillet features are defined for generating inspection features. They are all subtypes of SimpleFeature. Figure 11 shows the diagram of all the fillet feature classes in the FilletFeaturePack subpackage.

Class Chamfer is used to represent a chamfer of an edge in a workpiece. It has five attributes. Attribute *theFirstReferencePlane* represents the first reference plane in creating a chamfer, and its data type is DatumPlane. Attribute *firstOffset* represents the first offset from the first reference plane for creating a chamfer, and its data type is LengthMeasure. Attribute *theSecondReferencePlane* represents the second reference plane in creating a chamfer, and its data type is DatumPlane. Attribute *secondOffset* represents the second offset from the second reference plane for creating a chamfer, and its data type is LengthMeasure. Attribute *thePlane* represents the plane of the chamfer, and its data type is PlaneFeature.

Class Fillet is used to represent a fillet of an edge in a workpiece. It has two attributes. Attribute *edge* represents the edge on which a fillet is placed, and its data type is GCurveFeature3D. Attribute *radius* represents the radius of the fillet, and its data type is LengthMeasure.

29

Class RoundEdge is used to represent a rounded edge in a workpiece. It has two attributes. Attribute *edge* represents the edge that a fillet is placed, and its data type is GCurveFeature3D. Attribute *radius* represents the radius of the fillet, and its data type is LengthMeasure.

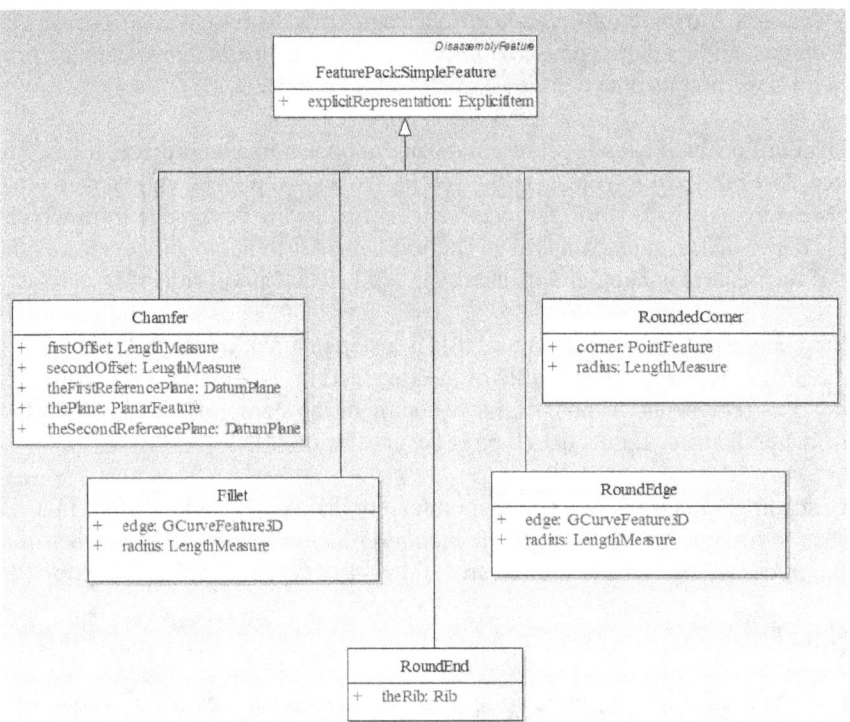

Figure 11. Class diagram of FilletFeaturePack

Class RoundedCorner is used to represent a rounded corner in a workpiece. It has two attributes. Attribute *corner* represents the corner that is rounded, and its data type is PointFeature. Attribute *radius* represents the radius of the round corner, and its data type is LengthMeasure.

3.2.5 Special Features

Special features are defined for generating features with special purposes, such as thread and screw. More special features, such as gear tooth and helical spring, can be added when they are in need. Figure 12 shows the diagram of all the special feature classes in the SpecialFeaturePack package.

Class GeneralThread is used to represent a general thread. It has eight attributes. Attribute *fitClass* represents how tight the assembled threads are, and its data type is String. Fit class is defined in ISO 10303-Part 224. Attribute *form* represents the geometric shape of the thread, and its data type is String. Attribute *innerThread*

30

represents whether the thread is an inner thread. If it is not, it is an outer thread. The data type is **boolean**. Attribute *majorDiameter* represents the major diameter of the thread, and its data type is LengthMeasure. Attribute *minorDiameter* represents the minor diameter of the thread, and its data type is LengthMeasure. Attribute *pitchDiameter* represents the pitch diameter of the thread, and its data type is LengthMeasure. Attribute *numberOfThread* represents the number of threads in a unit of length, and its data type is **int**. Attribute *rightHanded* represents whether the thread is right handed in orientation. If it is not, it is left handed. The data type is **boolean**.

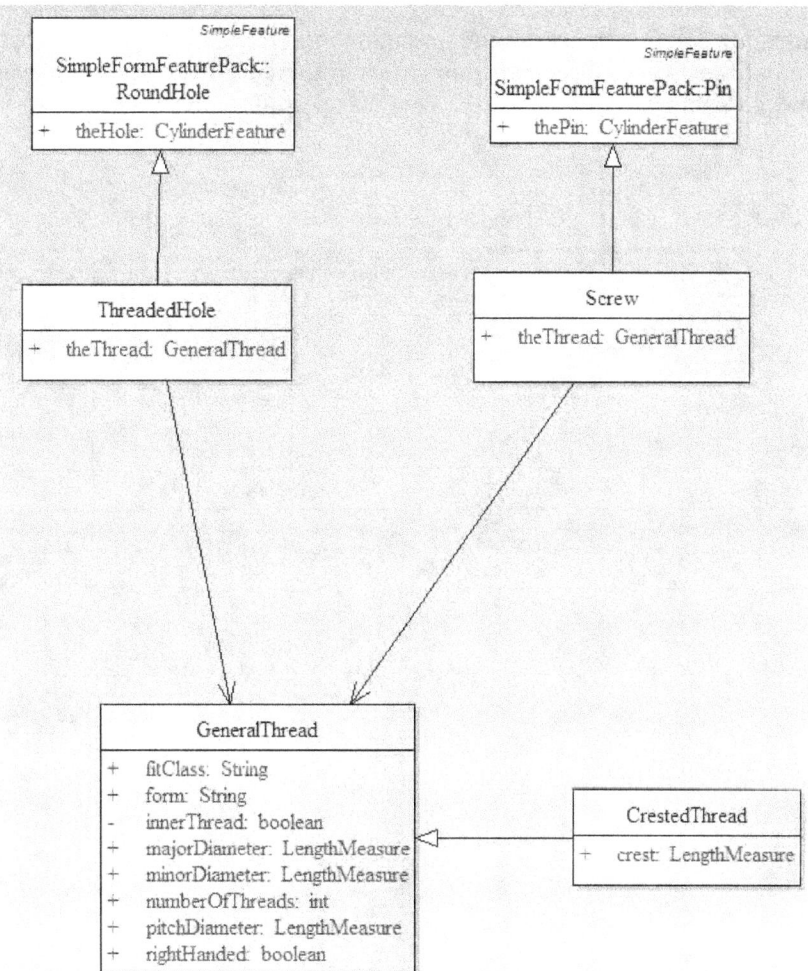

Figure 12. Class diagram of SpecialFeaturePack

Class ThreadedHole is a subtype of RoundHole and is used to represent a threaded hole in a workpiece. It has one attribute. Attribute *theThread* represents the thread of a hole, and its data type is GeneralThread.

Class Screw is a subtype of Pin and is used to represent a threaded pin in a workpiece. It has one attribute. Attribute *theThread* represents the thread of a pin, and its data type is GeneralThread.

Class CrestedThread is a subtype of GeneralThread and is used to represent a thread with a crest. It has one attribute. Attribute *crest* represents the dimension of the crest, and its data type is LengthMeasure.

3.2.6 Composite Form Features

Composite form features are used for representing form features that are composed of simple form features to create complex form features. Figure 13 shows the diagram of all the composite form feature classes in the CompositeFormFeaturePack package.

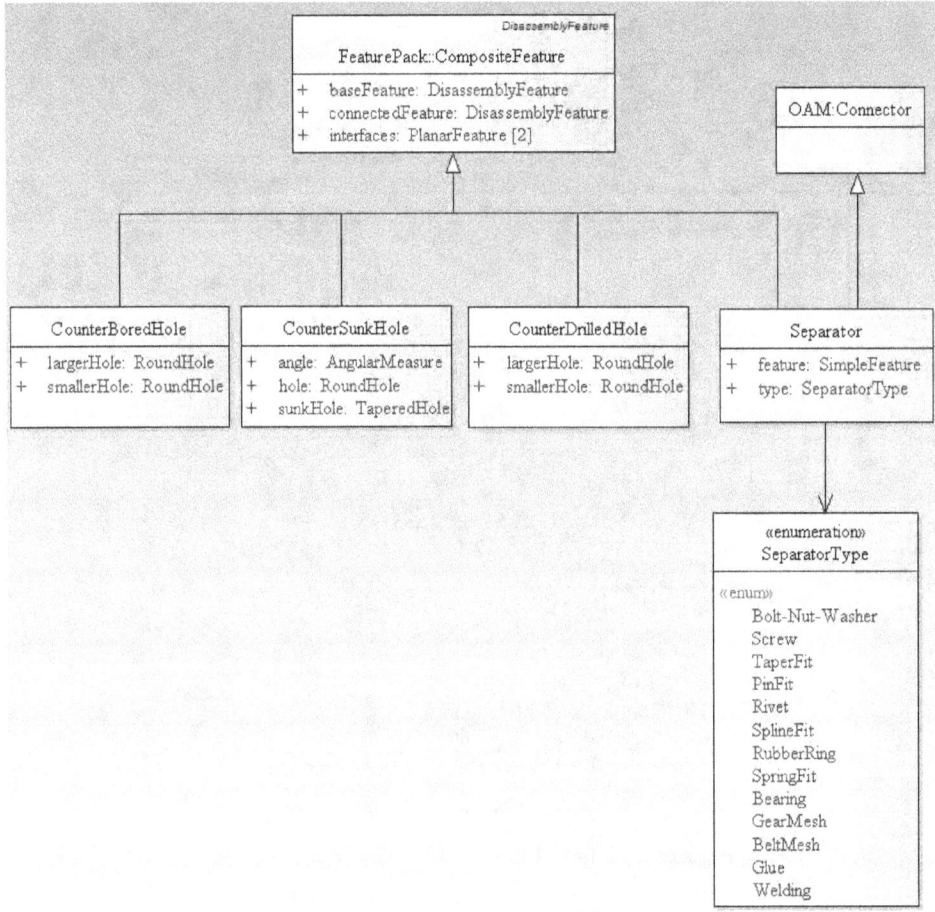

Figure 13. Class Diagram of CompositeFormFeaturePack

Class CounterBoredHole is used to represent a counterbored hole in a workpiece. It has two attributes. Attribute *largeHole* represents the larger hole in the counterbored hole, and its data type is RoundHole. Attribute *smallHole* represents the smaller hole in the counterbored hole, and its data type is RoundHole.

Class CounterSunkHole is used to represent a countersunk hole in a workpiece. It has three attributes. Attribute *hole* represents the hole in the countersunk hole, and its data type is RoundHole. Attribute *sunkHole* represents the tapered hole in the counter sunk hole, and its data type is TaperedHole. Attribute *angle* represents the taper angle, and its data type is AngularMeasure.

Class CounterDrilledHole is used to represent a counter drilled hole in a workpiece. It has two attributes. Attribute *largeHole* represents the larger hole in the counter drilled hole, and its data type is RoundHole. Attribute *smallHole* represents the smaller hole in the counter drilled hole, and its data type is RoundHole.

Enumeration type SeparatorType is used to define the type of a separator so that proper disassembly tools can be selected. This enumeration type includes *Bolt-Nut-Washer*, *Screw*, *TaperFit*, *PinFit*, *Rivet*, *SplineFit*, *RubberRing*, *SpringFit*, *Bearing*, *GearMesh*, *BeltMesh*, *Glue*, and *Welding*. Figure 14 shows a diagram of these types [Tseng 1999].

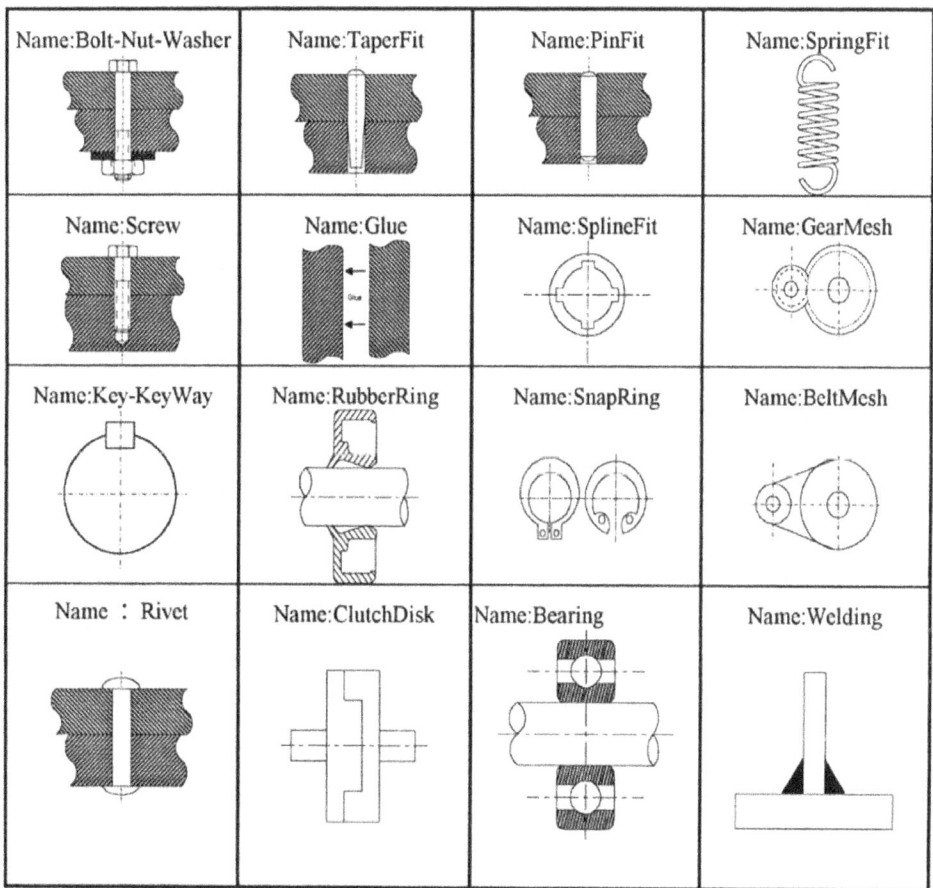

Figure 14. Types of separator

33

Class Separator is a subtype of Class Connector, defined in the Open Assembly Model, and is used to represent a separator in an assembly that is to be separated at the end of its useful life. It has two attributes. Attribute *feature* represents the feature from which the assembly will be separated into subassemblies, and its data type is SimpleFeature. Attribute *type* represents the type of the separator, and its data type is SeparatorType.

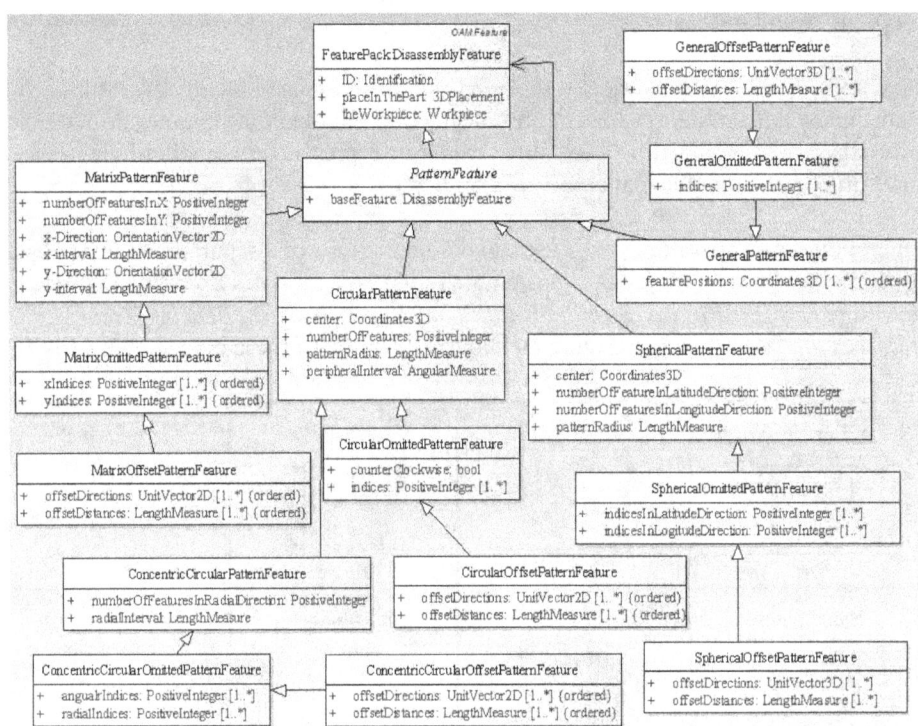

Figure 15. Class diagram of PatternFeaturePack

3.2.7 Pattern Features

Class PatternFeature is a subtype of Class DisassemblyFeature, and is used to represent a pattern of features in a workpiece. Figure 15 shows the diagram of all the pattern feature classes in the PatternFeaturePack package. Pattern feature is defined in ANSI/ASME Y14.5 standard [Y14.5 1994]. The class has one attribute. Attribute *baseFeature* represents a reference feature for all the features in the pattern, and its data type is Feature. The reference feature can be a simple feature or a composite feature.

Class MatrixPatternFeature is a subtype of Class PatternFeature, and is used to represent a matrix of features in a workpiece. The class has six attributes. Attribute *numberOfFeaturesInX* represents the number of features in the X direction of the pattern, and its data type is PositiveInteger. Attribute *numberOfFeaturesInY* represents the number of features in the Y direction of the pattern, and its data type is PositiveInteger.

34

Attribute *x-Direction* represents the X direction of the pattern that features form on a 2D plane, and its data type is OrientationVector2D. The origin of the vector must be placed on the hole that has index 1 in the X direction. Attribute *x-Interval* represents the interval between any two features in the X direction of the pattern of features, and its data type is LengthMeasure. Attribute *y-Direction* represents the Y direction of the pattern that features form on a 2D plane, and its data type is OrientationVector2D. The origin of the vector must be placed on the hole that has index 1 in the Y direction. Attribute *y-Interval* represents the interval between any two features in the Y direction of the pattern of features, and its data type is LengthMeasure.

Class MatrixOmittedPatternFeature is a subtype of Class MatrixPatternFeature, and is used to represent a matrix pattern of features with some features that are omitted in a workpiece. The class has two attributes. Attribute *xIndices* represents indices of the omitted features in the X direction of the pattern, and its data type is a set of PositiveInteger. Attribute *yIndices* represents indices of the omitted features in the Y direction of the pattern, and its data type is a set of PositiveInteger.

Class MatrixOffsetPatternFeature is a subtype of Class MatrixOmittedPatternFeature, and is used to represent a matrix pattern of features with some features that are offset from their normal positions. The class has two attributes. Attribute *offsetDirections* represents offset directions of the offset features, and its data type is a list of UnitVector2D. Attribute *offsetDistances* represents offset distances of the offset features, and its data type is a list of LengthMeasure.

Class CircularPatternFeature is a subtype of Class PatternFeature, and is used to represent a circular pattern of features in a workpiece. The class has five attributes. Attribute *numberOfFeatures* represents the number of features in the pattern, and its data type is PositiveInteger. Attribute *patternRadius* represents the radius of the pattern features, and its data type is LengthMeasure. Attribute *peripheralInterval* represents the interval between any two features in the tangential direction of the pattern of features, and its data type is AngularMeasure. Attribute *center* represents the center of the circular pattern, and its data type is Coordinates3D.

Class CircularOmittedPatternFeature is a subtype of Class CircularPatternFeature, and is used to represent a circular pattern of features with some features that are omitted in a workpiece. The class has two attributes. Attribute *counterClockwise* represents the sense of the indices of the omitted features in the tangential direction of the pattern, and its data type is **boolean**. The start point is on the positive X axis. Attribute *indices* represents indices of the omitted features in the pattern, and its data type is a set of PositiveInteger. The index starts from 1.

Class CircularOffsetPatternFeature is a subtype of Class CircularOmittedPatternFeature, and is used to represent a circular pattern of features with some features that are offset from their normal positions. The class has two attributes. Attribute *offsetDirections* represents offset directions of the offset features, and its data type is a list of

UnitVector2D. Attribute *offsetDistances* represents offset distances of the offset features, and its data type is a list of LengthMeasure.

Class ConcentricCircularPatternFeature is a subtype of Class CircularPatternFeature, and is used to represent a concentric circular pattern of features in a workpiece. The class has two attributes. Attribute *numberOfFeaturesInRadialDirection* represents the number of circular pattern features in the radial direction, and its data type is PositiveInteger. Attribute *radialInterval* represents the interval between any adjacent sets of circular features in the radial direction, and its data type is LengthMeasure.

Class ConcentricCircularOmittedPatternFeature is a subtype of Class ConcentricCircularPatternFeature, and is used to represent a concentric circular pattern of features with some features that are omitted in a workpiece. The class has two attributes. Attribute *angularIndices* represents indices of the omitted features in angular direction, and its data type is a set of PositiveInteger. Attribute *radialIndices* represents indices of the omitted features in radial direction, and its data type is a set of PositiveInteger.

Class ConcentricCircularOffsetPatternFeature is a subtype of Class ConcentricCircularOmittedPatternFeature, and is used to represent a concentric circular pattern of features with some features that are offset from their normal positions. The class has two attributes. Attribute *offsetDirections* represents offset directions of the offset features, and its data type is a list of UnitVector2D. Attribute *offsetDistances* represents offset distances of the offset features, and its data type is a list of LengthMeasure.

Class SphericalPatternFeature is a subtype of Class PatternFeature, and is used to represent a spherical pattern of features in a workpiece. The class has four attributes. Attribute *numberOfFeaturesInLatitudeDirection* represents the number of features in the latitude direction of the spherical pattern, and its data type is PositiveInteger. Attribute *numberOfFeaturesInLongitudeDirection* represents the number of features in the longitudinal direction of the spherical pattern, and its data type is PositiveInteger. Attribute *patternRadius* represents the radius of the features in the pattern of features, and its data type is LengthMeasure. Attribute *center* represents the center of the spherical pattern, and its data type is Coordintates3D.

Class SphericalOmittedPatternFeature is a subtype of Class SphericalPatternFeature, and is used to represent a spherical pattern of features with some features that are omitted in a workpiece. The class has two attributes. Attribute *indicesInLatitudeDirection* represents indices of the omitted features in the latitudinal direction, and its data type is a set of PositiveInteger. Attribute *indicesInLongitudeDirection* represents indices of the omitted features in the longitudinal direction, and its data type is a set of PositiveInteger.

Class SphericalOffsetPatternFeature is a subtype of Class SphericalOmittedPatternFeature, and is used to represent a spherical pattern of features with some features that are offset from their normal positions in a workpiece. The class has two attributes. Attribute *offsetDirections* represents offset directions of the offset features, and its data type is a list of UnitVector3D. Attribute *offsetDistances* represents offset distances of the offset features, and its data type is a list of LengthMeasure.

Class GeneralPatternFeature is a subtype of Class PatternFeature, and is used to represent a general pattern of features in a workpiece. The class has one attribute. Attribute *featurePositions* represents the positions of features in the pattern, and its data type is a list of Coordinates3D.

Class GeneralOmittedPatternFeature is a subtype of Class GeneralPatternFeature, and is used to represent a general pattern of features with some omitted features in the pattern. The class has one attribute. Attribute *indices* represents the indices of features in the pattern that are omitted, and its data type is a list of PositiveInteger.

Class GeneralOffsetPatternFeature is a subtype of Class GeneralOmittedPatternFeature, and is used to represent a general pattern of features with some offset features from their normal positions in the pattern. The class has two attributes. Attribute *offsetDirections* represents offset directions of the offset features, and its data type is a list of UnitVector3D. Attribute *offsetDistances* represents offset distances of the offset features, and its data type is a list of LengthMeasure.

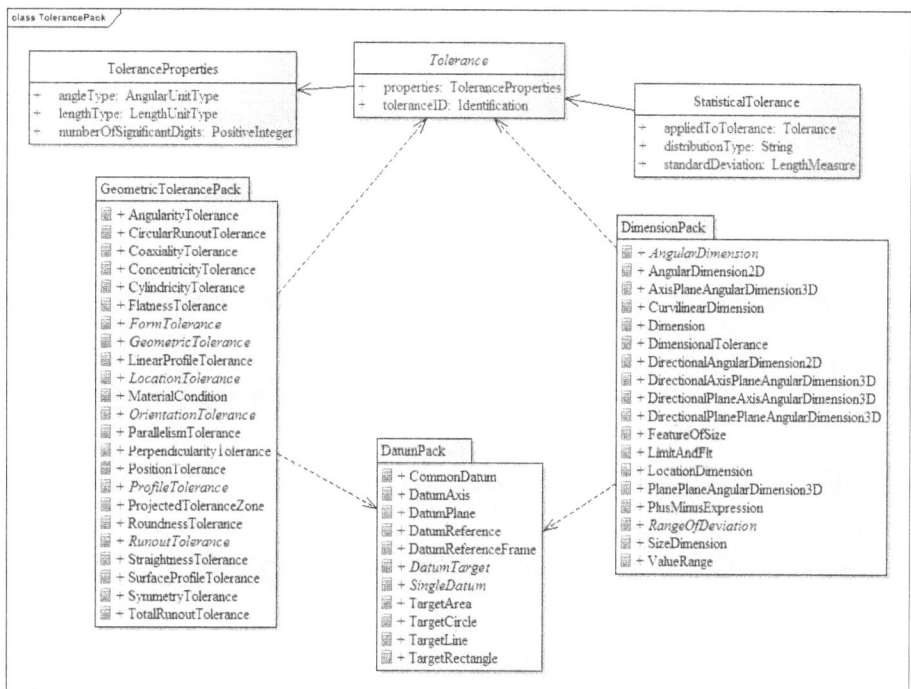

Figure 16. Class diagram of TolerancePack

3.3 Tolerance

Tolerance represents the allowable variation of a manufactured feature from the nominal design. Many concepts on datum establishing, dimensioning, and tolerancing are defined

37

in ANSI/ASME Y14.5. All the classes relevant to tolerancing are in a package called TolerancePack. It includes three top-level classes and three subpackages. Figure 16 shows the diagram of the TolerancePack package.

Class ToleranceProperties is used to represent properties of a tolerance. The class has three attributes. Attribute *angleType* represents the unit of angular measure used in the tolerance, and its data type is AngularUnitType. Attribute *lengthType* represents the unit of length measure used in the tolerance, and its data type is LengthUnitType. Attribute *numberOfSignificantDigits* represents the number of significant digits used in the tolerance, and its data type is PositiveInteger.

Class Tolerance is used to represent a generic tolerance. It is an abstract class. The class has two attributes. Attribute *properties* represents related properties of a tolerance, and its data type is ToleranceProperties. Attribute *toleranceID* represents the identification of the tolerance, and its data type is Identification.

Class StatisticalTolerance is used to represent a statistical expression in a tolerance. The class has three attributes. Attribute *appliedToTolerance* represents the tolerance to which statistical distribution and deviation are applied, and its data type is Tolerance. Attribute *distributionType* represents a statistical distribution, such as Gaussian or Binomial, and its data type is String. Attribute *standardDeviation* represents the standard deviation of the statistical distribution, and its data type is LengthMeasure.

Subpackages in the TolerancePack package include DatumPack, GeometricTolerancePack, and DimensionPack. They are described in the following subsections.

3.3.1 Datum

Datum represents a datum referenced in a tolerance or dimension. Figure 17 shows the diagram of all the classes in the DatumPack package.

Class Datum is used to represent a datum. It is an abstract class. The class has no attributes.

Class SingleDatum is used to represent a single datum feature. It is a subtype of *Datum*. The class has one attribute. Attribute *datumName* represents the name of the single datum, and the attribute's data type is String.

Class CommonDatum is used to represent a common datum. It is a subtype of *Datum*. The class has one attribute. Attribute *composedDatums* represents single datum features that are used to establish the common datum, and the attribute's data type is a set of two or more SingleDatum.

Class DatumReferenceFrame is used to represent a datum reference frame for a specified tolerance. The class has three attributes. Attribute *primaryDatum* represents the primary datum in the datum reference frame, and the attribute's data type is a Datum. Attribute *secondaryDatum* represents the secondary datum in the datum reference frame, and the attribute's data type is Datum. Attribute *tertiaryDatum* represents the tertiary datum in the datum reference frame, and the attribute's data type is Datum.

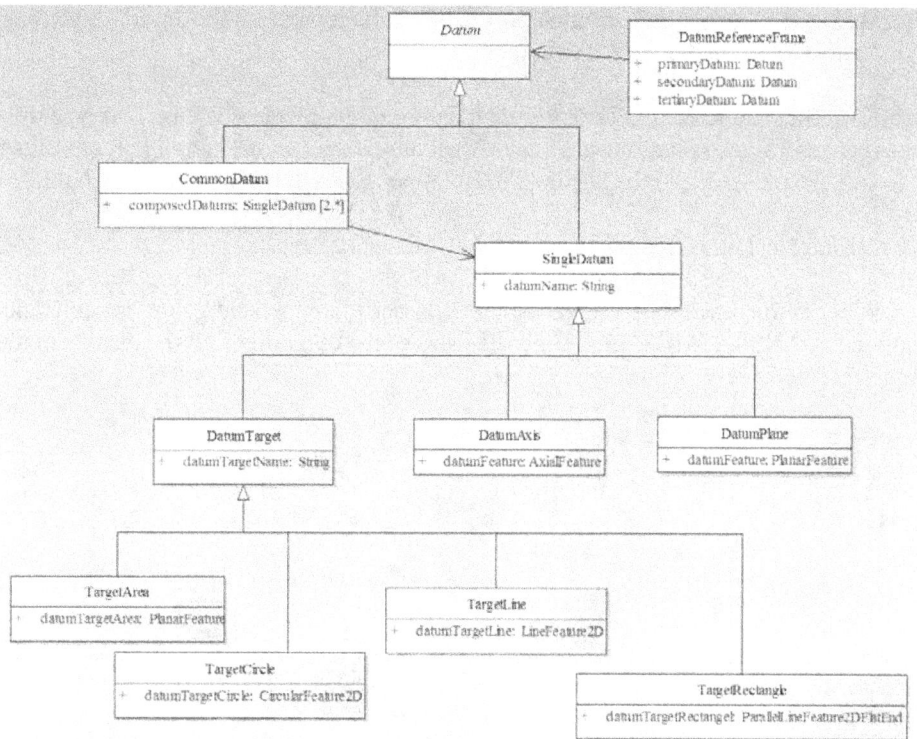

Figure 17. Class diagram of DatumPack

Class DatumTarget is used to represent a datum target. It is a subtype of SingleDatum. The class has one attribute. Attribute *datumTargetName* represents the name of the datum target, and the attribute's data type is String.

Class DatumAxis is used to represent a datum axis. It is a subtype of SingleDatum. The class has one attribute. Attribute *datumFeature* represents the axial feature that is used to define the datum axis, and the attribute's data type is AxialFeature.

Class DatumPlane is used to represent a datum plane. It is a subtype of SingleDatum. The class has one attribute. Attribute *datumFeature* represents the planar feature that is used to define the datum plane, and the attribute's data type is PlanarFeature.

Class TargetArea is used to represent a datum target area. It is a subtype of DatumTarget. The class has one attribute. Attribute *datumTargetArea* represents the planar feature that is used to define the datum target area, and the attribute's data type is PlanarFeature.

Class TargetCircle is used to represent a datum target circle. It is a subtype of DatumTarget. The class has one attribute. Attribute *datumTargetCircle* represents the datum target circle, and the attribute's data type is CircularFeature2D.

Class TargetLine is used to represent a datum target line. It is a subtype of DatumTarget. The class has one attribute. Attribute *datumTargetLine* represents the datum target line, and the attribute's data type is LineFeature2D.

Class TargetRectangle is used to represent a datum target rectangle. It is a subtype of DatumTarget. The class has one attribute. Attribute *datumTargetRectangle* represents the datum target rectangel, and the attribute's data type is ParallelLineFeature2DFlatEnd.

3.3.2 Geometric Tolerance

The geometric tolerance subpackage includes classes that represent geometric tolerances as defined in ANSI/ASME Y15.4. Figure 18 shows the diagram of all the classes in the GeometricTolerancePack package.

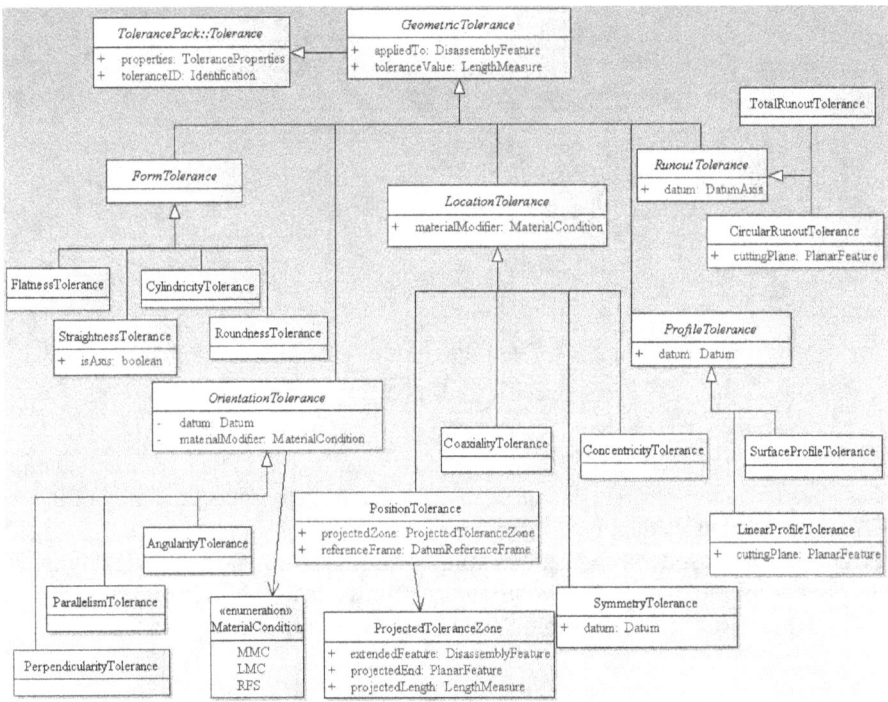

Figure 18. Class diagram of GeometricTolerancePack

Class *GeometricTolerance* is used to represent a generic geometric tolerance. It is a subtype of Tolerance and an abstract data type. The class has two attributes. Attribute *appliedTo* represents the feature to which a geometric tolerance is applied, and the attribute's data type is DisassemblyFeature. Attribute *toleranceValue* represents the magnitude of a tolerance, and the attribute's data type is LengthMeasure.

Class *FormTolerance* is used to represent a form tolerance. It is a subtype of *GeometricTolerance* and an abstract data type. The class has no additional attribute.

Class FlatnessTolerance is used to represent a flatness tolerance. It is a subtype of *FormTolerance*. The class has no additional attribute.

Class CylindricityTolerance is used to represent a cylindricity tolerance. It is a subtype of *FormTolerance*. The class has no additional attribute.

Class RoundnessTolerance is used to represent a roundness tolerance. It is a subtype of *FormTolerance*. The class has no additional attribute.

Class StraightnessTolerance is used to represent a straightness tolerance. It is a subtype of *FormTolerance*. The class has one attribute. Attribute *isAxis* represents whether the tolerance is applied to an axis, and the attribute's data type is **boolean**. If it is not applied to an axis, it is applied to a line or parallel lines on a planar feature.

Enumeration type MaterialCondition is used to define the material condition that the magnitude of a tolerance can change. This enumeration type includes *MMC, LMC*, and *RFS*. MMC standards for Maximum Material Condition. LMC standards for Least Material Condition. RFS standards for Regardless of Feature Size.

Class OrientationTolerance is used to represent an orientation tolerance. It is a subtype of *GeometricTolerance* and an abstract data type. The class has two attributes. Attribute *datum* represents the datum to which the tolerance is referenced, and the attribute's data type is Datum. Attribute *materialModifier* represents the material condition of the tolerance, and the attribute's data type is MaterialCondition.

Class PerpendicularityTolerance is used to represent a perpendicularity tolerance. It is a subtype of *OrientationTolerance*. The class has no additional attribute.

Class ParallelismTolerance is used to represent a parallelism tolerance. It is a subtype of *OrientationTolerance*. The class has no additional attribute.

Class AngularityTolerance is used to represent an angularity tolerance. It is a subtype of *OrientationTolerance*. The class has no additional attribute.

Class *LocationTolerance* is used to represent a location tolerance. It is a subtype of *GeometricTolerance* and an abstract data type. The class has one attribute. Attribute *materialModifier* represents the material condition of the location tolerance, and the attribute's data type is MaterialCondition.

Class ProjectedToleranceZone is used to represent a tolerance zone projected from a feature to which a tolerance is applied. The class has three attributes. Attribute *extendedFeature* represents the feature from which the tolerance zone is extended, and the attribute's data type is DisassemblyFeature. Attribute *projectedEnd* represents the end of the feature (a planar feature) to which the projected tolerance zone is connected, and the attribute's data type is PlanarFeature. Attribute *projectedLength* represents the length of the projected tolerance zone, and the attribute's data type is LengthMeasure.

Class PositionTolerance is used to represent a position tolerance. The class is a subtype of *LocationTolerance* and has two attributes. Attribute *projectedZone* represents a projected tolerance zone, and the attribute's data type is ProjectedToleranceZone. If the length of the project tolerance zone is zero, it means that there is no projected tolerance zone. Attribute *referenceFrame* represents the datum reference frame to which the position tolerance is referenced, and the attribute's data type is DatumReferenceFrame.

Class CoaxialityTolerance is used to represent a coaxiality tolerance. It is a subtype of *LocationTolerance*. The class has no additional attribute.

Class SymmetryTolerance is used to represent a symmetry tolerance. It is a subtype of *LocationTolerance*. The class has one attribute. Attribute *datum* represents the datum to which the symmetry tolerance is referenced, and the attribute's data type is Datum.

Class ConcentricityTolerance is used to represent a concentricity tolerance. It is a subtype of *LocationTolerance*. The class has no additional attribute.

Class *ProfileTolerance* is used to represent a profile tolerance. It is a subtype of *GeometricTolerance* and an abstract data type. The class has one attribute. Attribute *datum* represents the datum to which the profile tolerance is referenced, and the attribute's data type is Datum.

Class SurfaceProfileTolerance is used to represent a profile tolerance for a surface. It is a subtype of *ProfileTolerance*. The class has no additional attribute.

Class LinearProfileTolerance is used to represent the tolerance of a linear profile. It is a subtype of *ProfileTolerance*. The class has one attribute. Attribute *cuttingPlane* represents the cutting plane that defines the linear profile on a feature, and the attribute's data type is PlanarFeature.

Class *RunoutTolerance* is used to represent a runout tolerance. It is a subtype of *GeometricTolerance* and an abstract data type. The class has one attribute. Attribute *datum* represents the datum axis to which the runout tolerance is referenced, and the attribute's data type is DatumAxis.

Class CircularRunoutTolerance is used to represent a circular runout tolerance. It is a subtype of *RunoutTolerance*. The class has one attribute. Attribute *cuttingPlane* represents the cutting plane that defines a circle on a feature, and the attribute's data type is PlanarFeature.

Class TotalRunoutTolerance is used to represent a total runout tolerance. It is a subtype of *RunoutTolerance*. The class has no additional attribute.

3.3.3 Dimensional Tolerance

The dimensional tolerance subpackage includes classes that represent dimensional tolerances as defined in ANSI/ASME Y15.4. Figure 19 shows the diagram of all the classes in the DimensionalTolerancePack package.

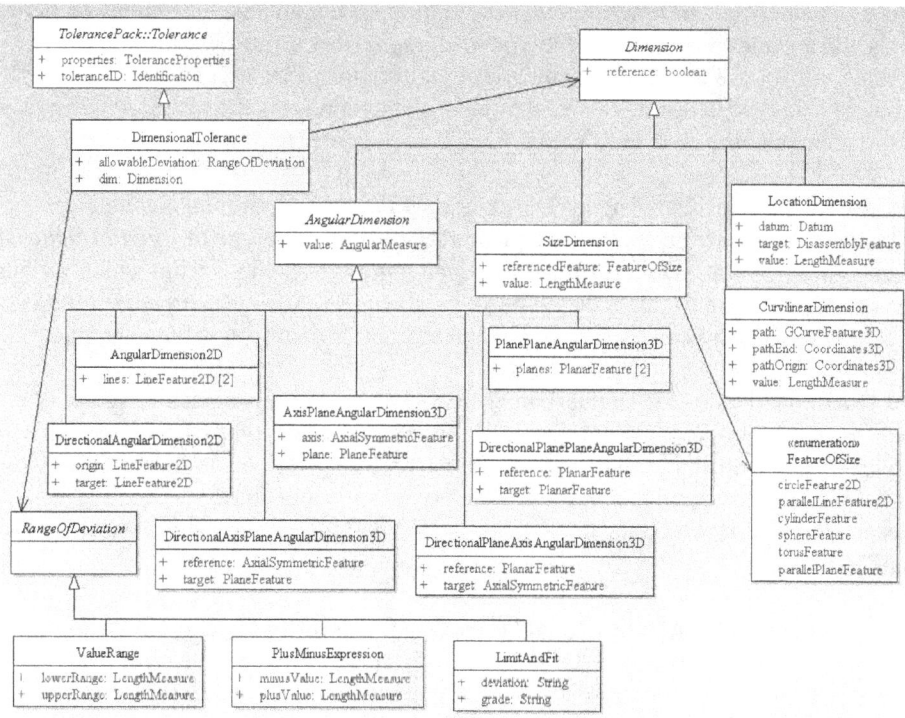

Figure 19. Class diagram of DimensionalTolerancePack

Class *Dimension* is used to represent a dimension of a feature. It is an abstract data type. The class has one attribute. Attribute *reference* represents whether the dimension is a reference dimension, and the attribute's data type is **boolean**. If it is not a reference dimension, then the dimension is a nominal dimension.

Class *AngularDimension* is used to represent an angular dimension of a feature. It is an abstract data type and a subtype of *Dimension*. The class has one attribute. Attribute *value* represents the value of the dimension, and the attribute's data type is AngularMeasure.

43

Class AngularDimension2D is used to represent an angular dimension of a feature. It is a subtype of *AngularDimension*. The class has one attribute. Attribute *lines* represents the two sides of the angular feature in measurement, and the attribute's data type is a set of two LineFeature2D.

Class DirectionalAngularDimension2D is used to represent a directional angular dimension of a feature. It is a subtype of *AngularDimension*. The class has two attributes. Attribute *origin* represents the baseline of the angular measurement, and the attribute's data type is LineFeature2D. Attribute *target* represents the target of the angular measurement, and the attribute's data type is LineFeature2D. The angular dimension is measured from the origin to the target.

Class AxisPlaneAngularDimension3D is used to represent an angular dimension between an axis and a plane feature. It is a subtype of *AngularDimension*. The class has two attributes. Attribute *axis* represents the axis in the angular measurement, and the attribute's data type is AxialFeature. Attribute *plane* represents the plane in the angular measurement, and the attribute's data type is PlaneFeature.

Class DirectionalAxisPlaneAngularDimension3D is used to represent an angular dimension measured from an axis to a plane feature. It is a subtype of *AngularDimension*. The class has two attributes. Attribute *reference* represents the axis from which the angle is measured, and the attribute's data type is AxialFeature. Attribute *target* represents the plane to which the angle is measured, and the attribute's data type is PlaneFeature.

Class DirectionalPlaneAxisAngularDimension3D is used to represent an angular dimension measured from a plane feature to an axis. It is a subtype of *AngularDimension*. The class has two attributes. Attribute *reference* represents the plane feature from which the angle is measured, and the attribute's data type is PlaneFeature. Attribute *target* represents the axis to which the angle is measured, and the attribute's data type is AxialFeature.

Class PlanePlaneAngularDimension3D is used to represent an angular dimension between two plane features. It is a subtype of *AngularDimension*. The class has one attribute. Attribute *planes* represents the two plane features in the angular measurement, and the attribute's data type is a set of two PlanarFeature.

Class DirectionalPlanePlaneAngularDimension3D is used to represent an angular dimension between two plane features. It is a subtype of *AngularDimension*. The class has two attributes. Attribute *reference* represents the plane feature from which the angle is measured, and the attribute's data type is PlanarFeature. Attribute *target* represents the plane to which the angle is measured, and the attribute's data type is PlanarFeature.

Enumeration type FeatureOfSize is used to list the features of size. Feature of size is described in Y14.5. This enumeration type includes *circleFeature2D, parallelLineFeature2D, cylinderFeature, sphereFeature, torusFeature,* and *parallelPlaneFeature.*

Class SizeDimension is used to represent the size dimension of a feature of size. It is a subtype of *Dimension*. The class has two attributes. Attribute *referencedFeature* represents the feature of size, and the attribute's data type is FeatureOfSize. Attribute *value* represents the value of the size dimension, and the attribute's data type is LengthMeasure.

Class CurvilinearDimension is used to represent the dimension of a curve on a feature. It is a subtype of *Dimension*. The class has four attributes. Attribute *path* represents the path of which a curve is measured, and the attribute's data type is GCurveFeature3D. Attribute *pathOrigin* represents the starting point of the measurement path, and the attribute's data type is Coordinates3D. Attribute *pathEnd* represents the end point of the measurement path, and the attribute's data type is Coordinates3D. Attribute *value* represents the value of the curvilinear dimension, and the attribute's data type is LengthMeasure.

Class LocationDimension is used to represent the location dimension of a feature. It is a subtype of *Dimension*. The class has three attributes. Attribute *datum* represents the reference from where the location of a feature is measured, and the attribute's data type is Datum. Attribute *target* represents the feature whose location relative to a datum is measured, and the attribute's data type is DisassemblyFeature. Attribute *value* represents the value of the curvilinear dimension, and the attribute's data type is LengthMeasure.

3.4 Workpiece

All the classes relevant to workpiece are in the WorkpiecePack package. It includes classes that represent a workpiece to be disassembled for remanufacturing, reuse, or recycling. A workpiece to be disassembled can be the whole or a part of an end-of-service-life product. An end-of-service-life product to be remanufactured is also referred to as core. The disassembly process includes many subprocesses, such as separation, cleaning, and inspection. Figure 20 shows the diagram of classes in the WorkpiecePack package.

Class *PartCharacteristic* is used to represent a characteristic of a workpiece. It is an abstract data type. The class has no attributes.

Enumeration type SurfaceRoughnessType is used to list types of surface roughness. All the definitions on surface finish roughness can be found in ISO 1302 [ISO 1302 2002]. This enumeration type includes R_a (the arithmetic average of a set of absolute measured values) and R_{rms} (the root mean square of a set of measured values). The list can be extended when it is necessary.

Class SurfaceRoughness is used to represent the surface roughness of the feature of a workpiece that is being processed. It is a subtype of *PartCharacteristic*. The class has two attributes. Attribute *roughnessType* represents the type of the surface roughness, and the attribute's data type is SurfaceRoughnessType. Attribute *value* represents the value of the surface roughness, and the attribute's data type is MeasureWithUnit.

Class Reflectivity is used to represent the reflectivity of the surface of a workpiece that is being processed. It is a subtype of *PartCharacteristic*. The class has one attribute. Attribute *value* represents the value of the surface reflectivity, and the attribute's data type is MeasureWithUnit.

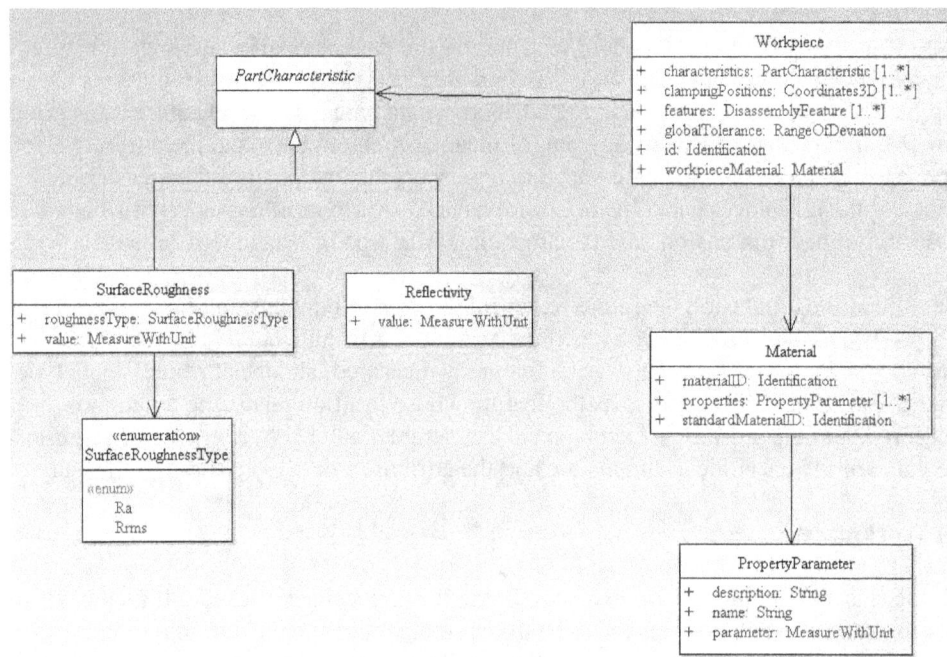

Figure 20. Class diagram of WorkpiecePack

Class PropertyParameter is used to represent the parameter of a workpiece material property. The class has three attributes. Attribute *description* represents the description of the material property parameter, and the attribute's data type is String. Attribute *name* represents the name of the material property parameter, and the attribute's data type is String. Attribute *parameter* represents the parameter of the material property, and the attribute's data type is MeasureWithUnit.

Class Material is used to represent the material of a workpiece. The class has three attributes. Attribute *materialID* represents the identification of the material, and the attribute's data type is Identification. Attribute *properties* represents the material properties, and the attribute's data type is a set of PropertyParameter. Attribute *standardMaterialID* represents the standard material identification, and the attribute's data type is Identification.

Class Workpiece is used to represent a workpiece. The class has six attributes. Attribute *characteristics* represents characteristics of the workpiece, and the attribute's data type is a set of *PartCharacteristic*. Attribute *clampingPositions* represents a set of clamping positions of the workpiece on a machine tool, and the attribute's data type is a set of

46

Coordinates3D. The clamping positions are in the machine tool coordinate system. Attribute *globalTolerance* represents the default tolerance that is used in dimensions of the workpiece, and the attribute's data type is RangeOfDeviation. Attribute *id* represents the identification of the workpiece, and the attribute's data type is Identification. Attribute *workpieceMaterial* represents the material of the workpiece, and the attribute's data type is Material. Attribute *features* represents a set of disassembly features in the workpiece, and the attribute's data type is a set of DisassemblyFeature.

3.5 Equipment

All the classes relevant to equipment are in the EquipmentPack package. It includes one class and three subpackages. Figure 21 shows the diagram of classes in the EquipmentPack package.

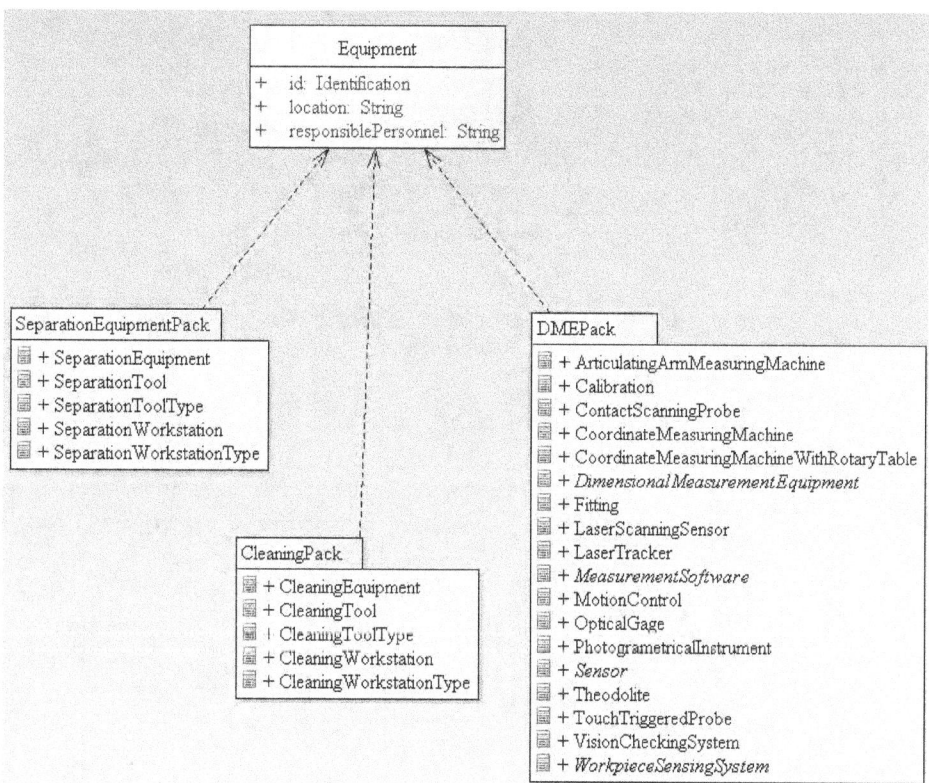

Figure 21. Class diagram of EquipmentPack

Class Equipment is used to represent a piece of equipment used in disassembly for reuse, recycle, or remanufacturing. The class has three attributes. Attribute *id* represents the identification of a piece of equipment, and the attribute's data type is Identification. Attribute *location* represents the location of the piece of equipment in a factory, and the attribute's data type is String. Attribute *responsiblePersonnel* represents the name of the person who is responsible for the equipment in a factory, and the attribute's data type is String.

Three subpackages are SeparationEquipmentPack, CleaningPack, and DMEPack. They are described in the following subsessions.

3.5.1 Separation Equipment

All the classes relevant to separation equipment are in the SeparationEquipmentPack package. It includes classes that represent a piece of equipment used in separating an assembly into subassemblies or individual parts. Figure 22 shows the diagram of classes in the package.

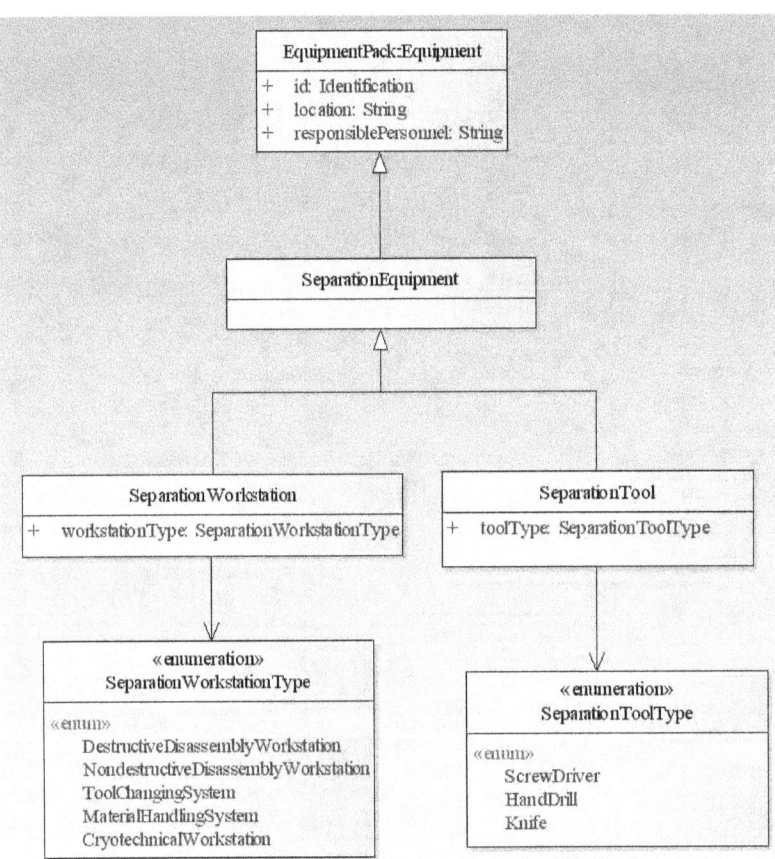

Figure 22. Class diagram of SeparationEquipmentPack

Class SeparationEquipment is used to represent a piece of equipment used in separating an assembly into subassemblies or individual parts for reuse, recycle, or remanufacturing. The class is a subtype of Equipment and has no additional attribute.

Enumeration type SeparationToolType is used to list types of tool used in separation. This enumeration type includes *ScrewDriver*, *HandDrill*, and *Knife*. The list can be extended when it is necessary.

Class SeparationTool is used to represent a tool used in separating an assembly for reuse, recycle, or remanufacturing. The class is a subtype of SeparationEquipment and has one attribute. Attribute *toolType* represents the type of a separation tool, and the attribute's data type is SeparationToolType.

Enumeration type SeparationWorkstationType is used to list types of workstation used in separation. This enumeration type includes *DestructiveDisassemblyWorkstation*, *NondestructiveDisassemblyWorkstation*, *ToolChangingSystem*, *MaterialHandlingSystem*, and *CryotechnicalWorkstation*. The list can be extended when it is necessary.

Class SeparationWorkstation is used to represent a workstation used in separating an assembly into subassemblies or individual parts for reuse, recycle, or remanufacturing. The class is a subtype of SeparationEquipment and has one attribute. Attribute *workstationType* represents the type of a separation workstation, and the attribute's data type is SeparationWorkstationType.

3.5.2 Cleaning Equipment

All the classes relevant to cleaning equipment are in the CleaningPack package. It includes classes that represent a piece of equipment used in cleaning a subassembly or an individual part. Figure 23 shows the diagram of classes in the package.

Class CleaningEquipment is used to represent a piece of equipment used in cleaning separated parts for reuse, recycle, or remanufacturing. The class is a subtype of Equipment and has no additional attribute.

Enumeration type CleaningWorkstationType is used to list types of workstation used in a cleaning process. This enumeration type includes *CompressedAirWorkstation*, *DryIceCleaningWorkstation*, *CO2-SnowWorkstation*, and *LaserCleaningWorkstation*. The list can be extended when it is necessary.

Enumeration type CleaningToolType is used to list types of tool used in cleaning. This enumeration type includes *CompressedAirSupply*. The list can be extended when it is necessary.

Class CleaningWorkstation is used to represent a workstation used in cleaning separated parts. The class is a subtype of CleaningEquipment and has one attribute. Attribute *type* represents the type of a separation workstation, and the attribute's data type is CleaningWorkstationType.

Class CleaningTool is used to represent a tool used in cleaning separated parts. The class is a subtype of CleaningEquipment and has one attribute. Attribute *type* represents the type of the cleaning tool, and the attribute's data type is CleaningToolType.

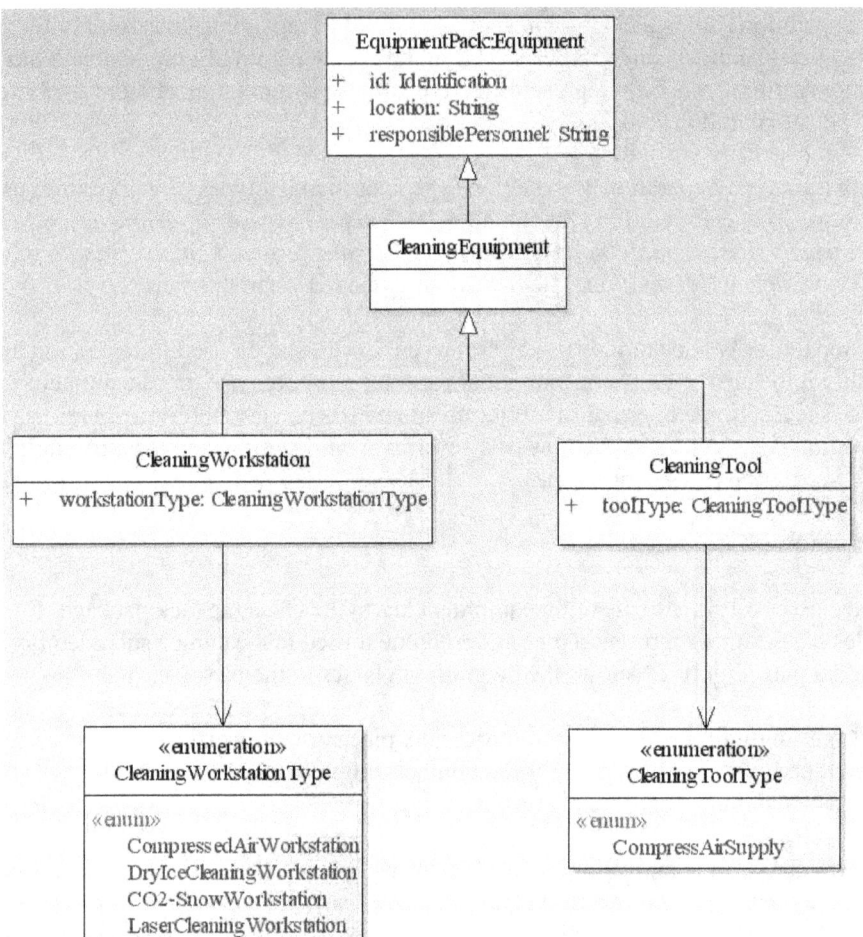

Figure 23. Class diagram of CleaningPack

3.5.3 Dimensional Measurement Equipment

All the classes relevant to dimensional measurement equipment are in the DMEPack package. It includes classes that represent a piece of equipment used in measuring a disassembled part. Figure 24 shows the diagram of classes in the package.

Class DimensionalMeasurementEquipment represents a piece of equipment used in dimensional measurement of separated parts. The class is a subtype of Equipment and has no additional attributes.

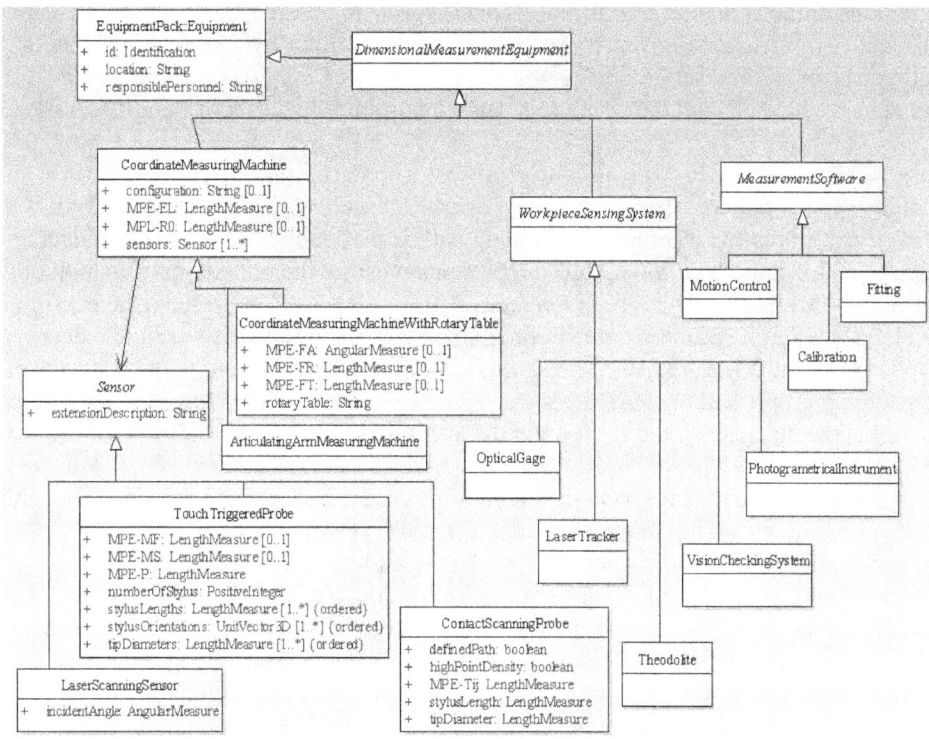

Figure 24. Class diagram of DMEPack

Class *Sensor* is used to represent a sensor used in dimensional inspection of separated parts. It is an abstract class. The class has one attribute. Attribute *extensionDescription* represents the description of the extension with which the sensor is mounted on a coordinate measurement machine, and the attribute's data type is String. If there is no extension, the description would be stated as "none."

Class LaserScanningSensor is used to represent a laser sensor used in scanning separated parts. It is a subtype of *Sensor*. The class has one attribute. Attribute *incidentAngle* represents the incident angle of the laser beam relative to the workpiece coordinate system, and the attribute's data type is AngularMeasure.

Class TouchTriggeredProbe is used to represent a touch-triggered probe used in probing separated parts. It is a subtype of *Sensor*. The class has seven attributes. Attribute *MPE-MF* represents the maximum permissible form error in the fixed multiple-stylus probing system, as defined in ISO 10360-1 [ISO10360 2000], and the attribute's data type is optional LengthMeasure. Attribute *MPE-MS* represents the maximum permissible size error in the fixed multiple-stylus probing system, as defined in ISO 10360-1, and the attribute's data type is optional LengthMeasure. Attribute *MPE-P* represents the maximum permissible probing error as defined in ISO 10360-1, and the attribute's data type is LengthMeasure. Attribute *numberOfStylus* represents the number of styli on the

51

probe, and the attribute's data type is PositiveInteger. Attribute *stylusLengths* represents the lengths of the styli, and the attribute's data type is an ordered set of LengthMeasure. Attribute *stylusOrientations* represents the orientations of the styli, and the attribute's data type is an ordered set of UnitVector3D. Attribute *tipDiameters* represents the diameters of the styli, and the attribute's data type is an ordered set of LengthMeasure.

Class ContactScanningProbe is used to represent a contact scanning probe used in scanning separated parts. It is a subtype of *Sensor*. The class has five attributes. Attribute *definedPath* represents whether the scanning path is predefined. The attribute's data type is **boolean**. Attribute *highPointDensity* represents whether the point density is high or low, and the attribute's data type is **boolean**. Attribute *MPE-Tij* represents the maximum permissible scanning probe error as defined in ISO 10360-1, and the attribute's data type is optional LengthMeasure. Attribute *stylusLength* represents the length of the stylus of the scanning probe, and the attribute's data type is LengthMeasure. Attribute *tipDiameter* represents the diameter of the stylus, and the attribute's data type is LengthMeasure.

Class CoordinateMeasuringMachine is used to represent a coordinate measuring machine (CMM) used in measuring separated parts. It is a subtype of *DimensionalMeasurementEquipment*. The class has four attributes. Attribute *configuration* represents a machine configuration as defined in ANSI/ASME B89.1.12M [B89 1985]. The attribute's data type is optional String. Attribute *MPE-EL* represents the maximum permissible error of length measurement as defined in B89.4.10360.2 [B89 2008], and the attribute's data type is optional LengthMeasure. Attribute *MPE-R0* represents the repeatability range of the maximum permissible error of length measurement as defined in B89.4.10360.2, and the attribute's data type is optional LengthMeasure. Attribute *sensors* represents sensors loaded on a CMM, and the attribute's data type is a set of *Sensor*.

Class ArticulatingArmMeasuringMachine is used to represent an articulating arm measuring machine used in measuring separated parts. It is a subtype of CoordinateMeasuringMachine. The class has no additional attributes.

Class CoordinateMeasuringMachineWithRotaryTable is used to represent a coordinate measuring machine with a rotary table. It is a subtype of CoordinateMeasuringMachine. The class has four attributes. Attribute *MPE-FA* represents the maximum permissible axial error of the rotary table as defined in ISO 10360-1, and the attribute's data type is optional AngularMeasure. Attribute *MPE-FR* represents the maximum permissible radial error of the rotary table as defined in ISO 10360-1, and the attribute's data type is optional LengthMeasure. Attribute *MPE-FT* represents the maximum permissible tangential error of the rotary table as defined in ISO 10360-1, and the attribute's data type is optional LengthMeasure. Attribute *rotaryTable* represents a description of the rotary table, and the attribute's data type is String.

Class *WorkpieceSensingSystem* is used to represent a measuring system using sensing techniques. It is an abstract class and a subtype of *DimensionalMeasurementEquipment*. The class has no additional attributes.

Class OpticalGage is used to represent an optical gage. It is a subtype of *WorkpieceSensingSystem*. The class has no additional attributes.

Class LaserTracker is used to represent a laser tracker. It is a subtype of *WorkpieceSensingSystem*. The class has no additional attributes.

Class Theodolite is used to represent a Theodolite machine. It is a subtype of *WorkpieceSensingSystem*. The class has no additional attributes.

Class VisionCheckingSystem is used to represent a vision checking system. It is a subtype of *WorkpieceSensingSystem*. The class has no additional attributes.

Class PhotogrammetricalInstrument is used to represent a photogrammetric instrument. It is a subtype of *WorkpieceSensingSystem*. The class has no additional attributes.

Class *MeasurementSoftware* is used to represent a measurement software system. It is an abstract class and a subtype of *DimensionalMeasurementEquipment*. The class has no additional attributes.

Class MotionControl is used to represent a motion control software system. It is a subtype of *MeasurementSoftware*. The class has no additional attributes.

Class Calibration is used to represent a software system for calibration. It is a subtype of *MeasurementSoftware*. The class has no additional attributes.

Class Fitting is used to represent a fitting software system. Examples of fitting include least-square fitting and minimax fitting. It is a subtype of *MeasurementSoftware*. The class has no additional attributes.

3.6 Workflow

All the classes relevant to workflow are in the WorkflowPack package. It includes classes and two subpackages that represent workflow. Figure 25 shows the diagram of WorkflowPack.

Class PostCondition is used to represent the post condition of a decision, such as where to branch out or a succeeding operation. The class has one attribute. Attribute *processElementID* represents the identification of the succeeding process element, and the attribute's data type is Identification.

Class ProcessElement is used to represent an element in a process, such as an operation, a decision node, a joint node, a while loop, or an end node. The class has three attributes. Attribute *elementID* represents the identification of the process element, and the attribute's data type is Identification. Attribute *subElements* represents child process elements, and the attribute's data type is an optional set of ProcessElement. Attribute *successor* represents the successor of the process element, and the attribute's data type is PostCondition.

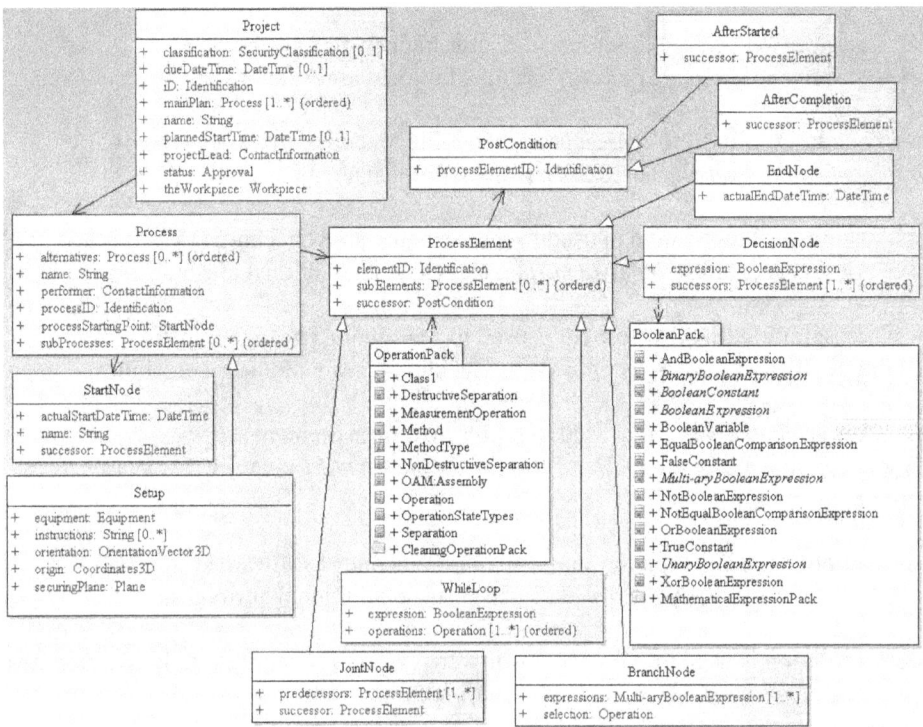

Figure 25. Class diagram of WorkflowPack

Class AfterStarted is used to represent a process element that occurs immediately after the process is started. (The start of this process can trigger the start of another process or other processes.) The class is a subtype of PostCondition and has one attribute. Attribute *successor* represents the succeeding process element, and the attribute's data type is ProcessElement.

Class AfterCompletion is used to represent a process element that occurs after the completion of an operation. The class is a subtype of PostCondition and has one attribute. Attribute *successor* represents the succeeding process element, and the attribute's data type is ProcessElement.

Class DecisionNode is used to represent a decision node in a process. Based on the result of a Boolean expression in the decision node, successors will be determined. The class has two attributes. Attribute *expression* represents the Boolean expression, and the attribute's data type is BooleanExpression. Class of Boolean expressions will be described in the subsection below. Attribute *successors* represents the succeeding process elements, and the attribute's data type is a list of ProcessElement.

Class JointNode is used to represent a joint node in a process. The class is a subtype of ProcessElement and has two attributes. Attribute *predecessors* represents the preceding process elements, and the attribute's data type is a set of ProcessElement. Attribute *successor* represents the succeeding process element, and the attribute's data type is ProcessElement.

Class WhileLoop is used to represent a while loop of process elements. The class is a subtype of ProcessElement and has two attributes. Attribute *expression* represents the Boolean expression of the termination of the while loop, and the attribute's data type is BooleanExpression. Attribute *operations* represents the operations within the while loop, and the attribute's data type is a list of Operation.

Class BranchNode is used to represent a selection of operation to be performed based on a predefined rule. The class is a subtype of ProcessElement and has two attributes. Attribute *expressions* is used to represent a set of rules defined by multiarity Boolean expressions, and the attribute's data type is a set of MultiArityBooleanExpression. Attribute *selection* represents the selected operation that satisfied the rules, and the attribute's data type is Operation.

Class StartNode is used to represent the start node of a process, such as disassembly or a subprocess, such as separation, cleaning, or inspection. The class has three attributes. Attribute *actualStartDateTime* represents the actual date and time of the start of a process, and the attribute's data type is DateTime. Attribute *name* represents the name of the process, and the attribute's data type is String. Attribute *successor* represents the successor of the start node, and the attribute's data type is ProcessElement.

Class Process is used to represent a process, consisting of a start node and process elements, including an end node. The class has six attributes. Attribute *name* represents the name of the process, and the attribute's data type is String. Attribute *performer* represents the performer of the process, and the attribute's data type is ContactInformation. Attribute *processID* represents the identification of the process, and the attribute's data type is Identification. Attribute *processStartingPoint* represents the start of the process, and the attribute's data type is StartNode. Attribute *subProcesses* represents child processes, and the attribute's data type is an optional list of ProcessElement. Attribute *alternatives* represents possible alternative processes, and the attribute's data type is an optional list of Process.

Class Setup is used to represent a setup process. The class is a subtype of Process and has five attributes. Attribute *equipment* represents the equipment used in the setup, and the attribute's data type is Equipment. Attribute *instructions* represents instructions of setting up, and the attribute's data type is an optional list of String. Attribute *orientation* represents the orientation of the workpiece in the machine coordinate system, and the attribute's data type is OrientationVector3D. Attribute *origin* represents the origin of the workpiece, and the attribute's data type is Coordinates3D. Attribute *securingPlane* represents a plane on the workpiece that is used to secure or clamp down the workpiece on a separation, cleaning, or inspection machine, and the attribute's data type is Plane.

Class Project is used to represent a project. The class has nine attributes. Attribute *dueDateTime* represents the project due date and time, and the attribute's data type is optional DateTime. Attribute *iD* represents the identification of the project, and the attribute's data type is Identification. Attribute *mainPlan* represents the main process plan of the project. The main process plan consists of one or more processes for disassembly. The attribute's data type is a list of Process. Attribute *name* represents the name of the project, and the attribute's data type is String. Attribute *plannedStartTime* represents the project starting date and time, and the attribute's data type is optional DateTime. Attribute *projectLead* represents the lead of the project, and the attribute's data type is ContactInformation. Attribute *status* represents the approval status of the project, and the attribute's data type is Approval.

The WorkflowPack includes two subpackages: BooleanPack and OperationPack. They are described in the following subsections.

3.6.1 Boolean Expression

All the classes relevant to Boolean Expression are in the BooleanPack package. It includes classes that represent Boolean constants, Boolean variables, binary Boolean expressions, multi-ary Boolean expressions, and mathematical expressions. They are used in determining the workflow based on conditions and predefined rules. Figure 26 shows the diagram of the BooleanPack subpackage.

Class *BooleanExpression* is used to represent a Boolean expression. The class is abstract and has no attribute.

Class *BooleanConstant* is used to represent a constant used in a Boolean expression. The class is a subtype of *BooleanExpression* and has no attribute.

Class TrueConstant is used to represent a Boolean constant of true. The class is a subtype of *BooleanConstant* and has no attribute.

Class FalseConstant is used to represent a Boolean constant of false. The class is a subtype of *BooleanConstant* and has no attribute.

Class BooleanVariable is used to represent a Boolean variable used in a Boolean expression. The class is a subtype of *BooleanExpression* and has two attributes. Attribute *name* represents the name of the variable, and the attribute's data type is String. Attribute *value* represents the value of the variable, and the attribute's data type is **boolean**.

Class NotBooleanExpression is used to represent a negation unary Boolean expression. The class is a subtype of *BooleanExpression* and has one attribute. Attribute *operand* represents the Boolean operand, and the attribute's data type is *BooleanExpression*.

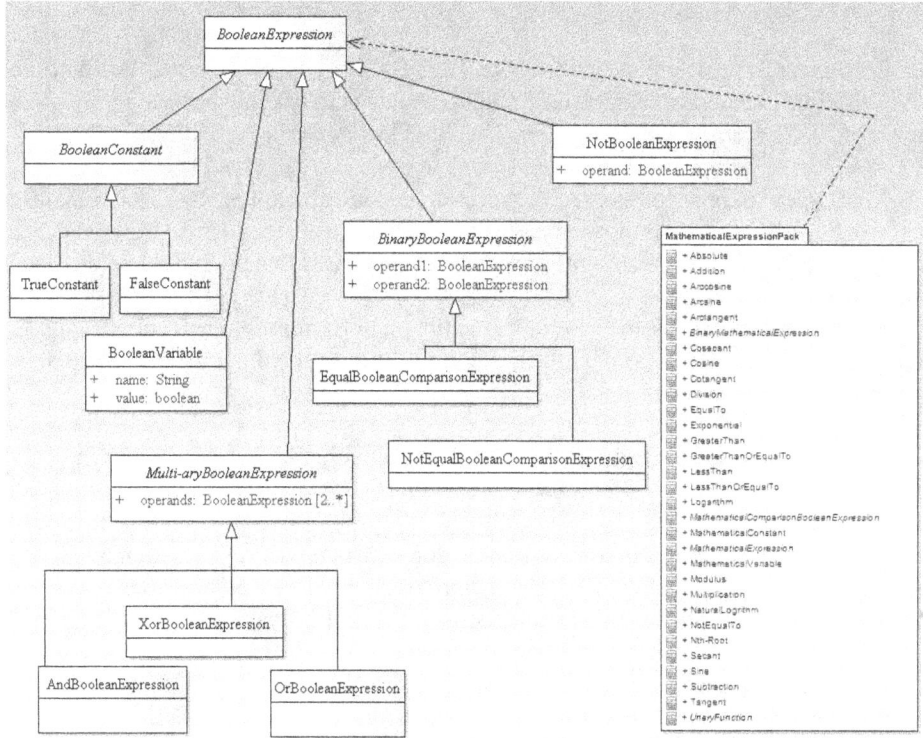

Figure 26. Class diagram of BooleanPack

Class *BinaryBooleanExpression* is used to represent a binary Boolean expression. The class is an abstract class and a subtype of *BooleanExpression*. The class has two attributes. Attribute *operand1* represents a Boolean operand, and the attribute's data type is *BooleanExpression*. Attribute *operand2* represents the other Boolean operand, and the attribute's data type is *BooleanExpression*.

Class EqualBooleanComparisonExpression is used to represent an evaluation of whether two operands are equal. The class is a subtype of *BinaryBooleanExpression*. The class has no additional attribute.

Class NotEqualBooleanComparisonExpression is used to represent an evaluation whether two operands are not equal. The class is a subtype of *BinaryBooleanExpression*. The class has no additional attribute.

Class Multi-aryBooleanExpression is used to represent a multi-ary Boolean expression. The class is an abstract class and a subtype of *BooleanExpression*. The class has one attribute. Attribute *operands* represents a set of two or more Boolean expressions, and the attribute's data type is a set of two or more *BooleanExpression*.

Class AndBooleanExpression is used to represent a Boolean expression that evaluates to be true if all of its operands evaluate to true and evaluates to false otherwise. The class is a subtype of *Multi-aryBooleanExpression*. The class has no additional attribute.

57

Class OrBooleanExpression is used to represent a Boolean expression that evaluates to false if all of its operands evaluate to false and evaluates to true otherwise. The class is a subtype of *Multi-aryBooleanExpression*. The class has no additional attribute.

Class XorBooleanExpression is used to represent a Boolean expression that evaluates to true if exactly one of its operands evaluates to true and evaluates to false otherwise. The class is a subtype of *Multi-aryBooleanExpression*. The class has no additional attribute.

All the mathematical expressions are in the MathematicExpressionPack subpackage, which is the subpackage of BooleanPack. Figure 27 shows the diagram of classes in MathematicExpressionPack.

Class *MathematicalExpression* is used to represent a mathematical expression. It is an abstract class and a subtype of *BooleanExpression*.

Class MathematicalConstant is used to represent a constant used in a mathematical expression. The class is a subtype of *MathematicalExpression*. The class has one attribute. Attribute *value* represents the value of the constant, and the attribute's data type is double.

Class MathematicalVariable is used to represent a variable used in a mathematical expression. It is a subtype of *MathematicalExpression* and has two attributes. Attribute *name* represents the name of the variable, and the attribute's data type is String. Attribute *value* represents the value of the variable, and the attribute's data type is MeasureWithUnit.

Class UnaryFunction is used to represent a mathematical function with only one operand. The class is a subtype of *MathematicalExpression*, and has one attribute. Attribute *operand* represents the operand used in the mathematical function, and the attribute's data type is *MathematicalExpression*.

Class Sine is used to represent a sine function. The class is a subtype of *UnaryFunction*. The class has no additional attribute.

Class Cosine is used to represent a cosine function. The class is a subtype of *UnaryFunction*. The class has no additional attribute.

Class Tangent is used to represent a tanget function. The class is a subtype of *UnaryFunction*. The class has no additional attribute.

Class Arctangent is used to represent an arctangent function. The class is a subtype of *UnaryFunction*. The class has no additional attribute.

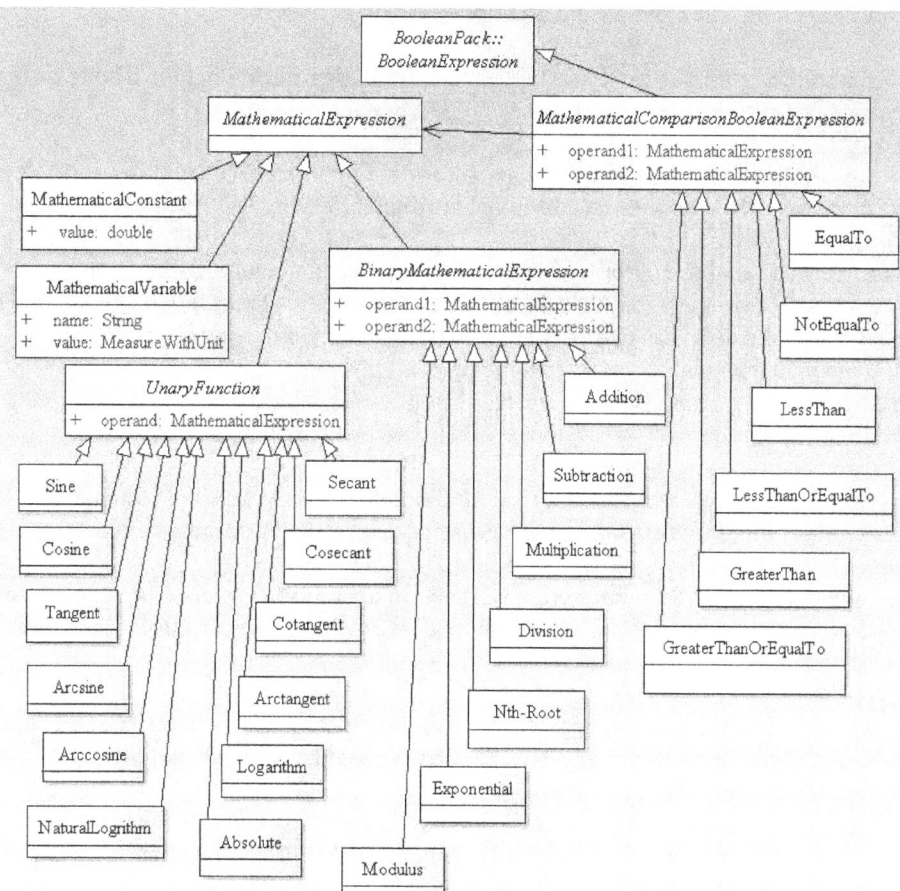

Figure 27. Class diagram of mathematical Comparison

Class Arcsine is used to represent an arcsine function. The class is a subtype of *UnaryFunction*. The class has no additional attribute.

Class Arccosine is used to represent an arccosine function. The class is a subtype of *UnaryFunction*. The class has no additional attribute.

Class Secant is used to represent a secant function. The class is a subtype of *UnaryFunction*. The class has no additional attribute.

Class Cosecant is used to represent a cosecant function. The class is a subtype of *UnaryFunction*. The class has no additional attribute.

Class Cotangent is used to represent a cotangent function. The class is a subtype of *UnaryFunction*. The class has no additional attribute.

Class Logarithm is used to represent a base 10 logarithm function. The class is a subtype of *UnaryFunction*. The class has no additional attribute.

Class NaturalLogarithm is used to represent a natural logarithm function. The class is a subtype of *UnaryFunction*. The class has no additional attribute.

Class Absolute is used to represent an absolute value function. The class is a subtype of *UnaryFunction*. The class has no additional attribute.

Class *BinaryMathematicalExpression* is used to represent a binary mathematical expression with two operands. The class is a subtype of *MathematicalExpression* and an abstract class. The class has two attributes. Attribute *operand1* represents an operand used in the mathematical function, and the attribute's data type is *MathematicalExpression*. Attribute *operand2* represents the other operand used in the mathematical function, and the attribute's data type is *MathematicalExpression*.

Class Addition is used to represent an addition operation. The class is a subtype of *BinaryMathematicalExpression*. The class has no additional attribute.

Class Subtraction is used to represent a subtraction operation. The class is a subtype of *BinaryMathematicalExpression*. The attribute *operand2* is subtracted from the *operand1*. The class has no additional attribute.

Class Multiplication is used to represent a multiplication operation. The class is a subtype of *BinaryMathematicalExpression*. The class has no additional attribute.

Class Division is used to represent a division operation. The class is a subtype of *BinaryMathematicalExpression*. The attribute *operand1* is divided by the *operand2*. The class has no additional attribute.

Class Nth-Root is used to represent an nth-root operation. The class is a subtype of *BinaryMathematicalExpression*. The attribute *operand1* is the base. The attribute *operand2* is the factor to perform the base in the nth-root operation. The class has no additional attribute.

Class Exponential is used to represent an exponential operation. The class is a subtype of *BinaryMathematicalExpression*. The attribute *operand1* is the base. The attribute *operand2* is the exponent. The class has no additional attribute.

Class Modulus is used to represent a modulus operation. The class is a subtype of *BinaryMathematicalExpression*. This operator returns the remainder when *operand1* is divided by *operand2*. The class has no additional attribute.

Class *MathematicalComparisonBooleanExpression* is used to represent a mathematical comparison expression of the two operands. The class is a subtype of *BooleanExpression* and an abstract class. The class has two attributes. Attribute *operand1* represents an operand used in the mathematical comparison, and the attribute's data type is *MathematicalExpression*. Attribute *operand2* represents the other operand used in the mathematical comparison, and the attribute's data type is *MathematicalExpression*.

Class EqualTo is used to represent an evaluation of whether *operand1* is equal to *operand2*. The class is a subtype of *MathematicalComparisonBooleanExpression*. The class has no additional attribute.

Class NotEqualTo is used to represent an evaluation of whether *operand1* is not equal to *operand2*. The class is a subtype of *MathematicalComparisonBooleanExpression*. The class has no additional attribute.

Class LessThan is used to represent an evaluation of whether *operand1* is less than *operand2*. The class is a subtype of *MathematicalComparisonBooleanExpression* . The class has no additional attribute.

Class LessThanOrEqualTo is used to represent an evaluation of whether *operand1* is less than or equal to *operand2*. The class is a subtype of *MathematicalComparisonBooleanExpression*. The class has no additional attribute.

Class GreaterThan is used to represent an evaluation of whether *operand1* is greater than *operand2*. The class is a subtype of *MathematicalComparisonBooleanExpression*. The class has no additional attribute.

Class GreaterThanOrEqualTo is used to represent an evaluation of whether *operand1* is greater than or equal to *operand2*. The class is a subtype of *MathematicalComparisonBooleanExpression*. The class has no additional attribute.

3.6.2 Operation

The OperationPack package includes classes that represent operations in disassembly, such as separation, cleaning, and dimensional inspection. Figure 28 shows the diagram of OperationPack, including a subpackage on the cleaning operation.

Enumeration type OperationStateTypes is used to list states of an operation. The states in the list include active, suspended, stopped, and resumed.

Class Operation is used to represent an operation in a process. The class is a subtype of ProcessElement and has four attributes. Attribute *alternatives* represents possible alternative operations, and the attribute's data type is an optional list of Operation. Attribute *onFeature* represents the feature on which the operation is performed, and the attribute's data type is DisassemblyFeature. Attribute *performer* represents the performer of the operation, and the attribute's data type is ContactInformation. Attribute *postCon* represents the post condition of the operation, and the attribute's data type is PostCondition. Attribute *state* represents the state of the operation, and the attribute's data type is OperationStateTypes.

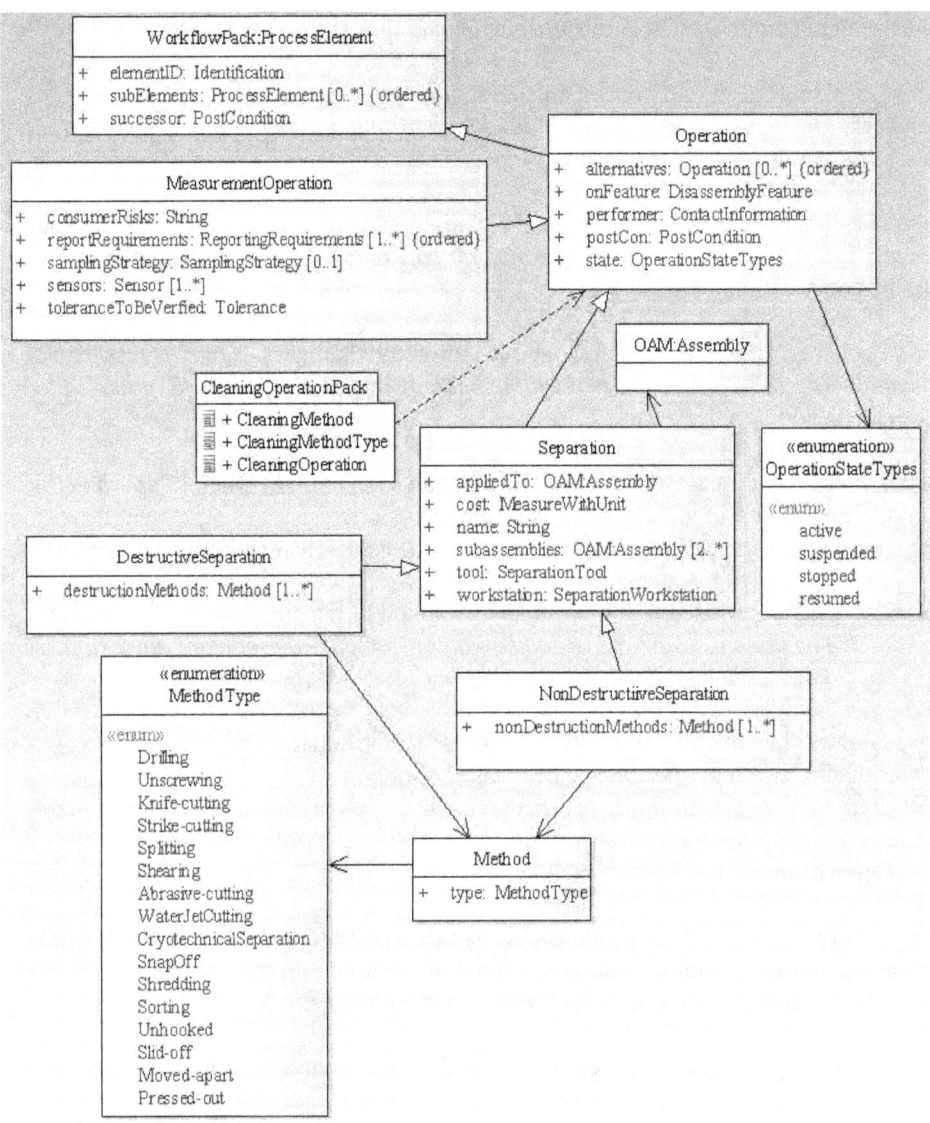

Figure 28. Class diagram of OperationPack

Class MeasurementOperation is used to represent a dimensional measurement operation. The class is a subtype of Operation and has five attributes. Attribute *consumerRisks* represents a description of the consumer risks that are associated with the measurement results, and the attribute's data type is String. Attribute *reportRequirements* represents the requirements of reporting measurement results, and the attribute's data type is a list of ReportingRequirement. Attribute *samplingStrategy* represents any specified sampling strategy associated with the measurement operation, and the attribute's data type is optional SamplingStrategy. Attribute *sensors* is used to represent sensors used in dimensional measurement of a feature, and the attribute's data type is a set of *Sensor*. Attribute *toleranceToBeVerified* represents the tolerance to be verified, and the attribute's data type is Tolerance.

Class Separation is used to represent a separation operation. The class is a subtype of Operation and has six attributes. Attribute *appliedTo* represents the assembly to which the separation operation is applied, and the attribute's data type is OAM:Assembly. OAM:Assembly is the assembly class of the Open Assembly Model (OAM) [Sudarsan 2003]. Attribute *cost* represents the cost that is associated with the separation, and the attribute's data type is MeasureWithUnit. Attribute *name* represents the name of the operation, and the attribute's data type is String. Attribute *subassemblies* represents the separated subassemblies as the result of the separation, and the attribute's data type is a set of OAM:Assembly. Attribute *tool* represents the tool used in the separation operation, and the attribute's data type is SeparationTool. Attribute *workstation* represents the workstation used in the separation operation, and the attribute's data type is SeparationWorkstation.

Enumeration type MethodType is used to list types of separation operations. This enumeration type includes *Drilling*, *Unscrewing*, *Knife-cutting*, *Strike-cutting*, *Splitting*, *Shearing*, *Abrasive-cutting*, *WaterJetCutting*, *CryotechnicalSeparation*, *SnapOff*, *Shredding*, *Sorting*, *Unhooked*, *Slid-off*, *Moved-apart*, and *Pressed-out*. The list can be extended when it is necessary.

Class Method is used to represent the method used in a separation operation. The class has one attribute. Attribute *type* represents the type of method, and the attribute's data type is MethodType.

Class DestructiveSeparation is used to represent a destructive separation operation. The class is a subtype of Separation and has one attribute. Attribute *destructiveMethods* represents methods used in a destructive separation, and the attribute's data type is a set of Method.

Class NonDestructiveSeparation is used to represent a nondestructive separation operation. The class is a subtype of Separation and has one attribute. Attribute *nonDestructiveMethods* represents methods used in a nondestructive separation, and the attribute's data type is a set of Method.

3.6.3 Cleaning Operation

The CleaningOperationPack package includes classes that represent cleaning operations in disassembly. Figure 29 shows the diagram of classes in the package.

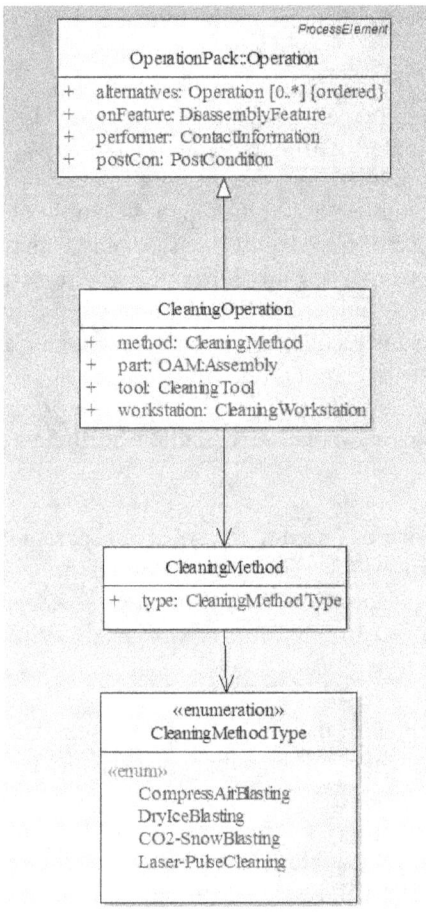

Figure 29. Class diagram of CleaningOperationPack

Enumeration type CleaningMethodType is used to list types of cleaning operations. This enumeration type includes *CompressAirBlasting, DryIceBlasting, CO2-SnowBlasting*, and *Laser-PulseCleaning*. The list can be extended when it is necessary.

Class CleaningMethod is used to represent the method used in a cleaning operation. The class has one attribute. Attribute *type* represents the type of method, and the attribute's data type is CleaningMethodType.

Class CleaningOperation is used to represent a cleaning operation. The class is a subtype of Operation and has four attributes. Attribute *method* represents a chosen cleaning method used in a cleaning operation, and the attribute's data type is a set of CleaningMethod. Attribute *part* represents the part or subassembly to which the cleaing operation is applied, and the attribute's data type is OAM:Assembly. Attribute *tool* represents the tool used in the cleaning operation, and the attribute's data type is

64

CleaningTool. Attribute *workstation* represents the workstation used in the cleaning operation, and the attribute's data type is CleaningWorkstation.

4. Case studies

A car suspension module and a flip-top cell phone are used to illustrate the use of the disassembly information model.

4.1. Car Suspension module

The car suspension module is assumed to be composed of four parts (*1-4*) and two sub-assemblies (*A* and *B*), as shown in Figure 30. The sub-assembly *A* contains two parts (*5* and *6*), and the sub-assembly *B* contains five parts (*7-11*) and a sub-assembly (*C*), which further decomposes into five parts (*12-16*).

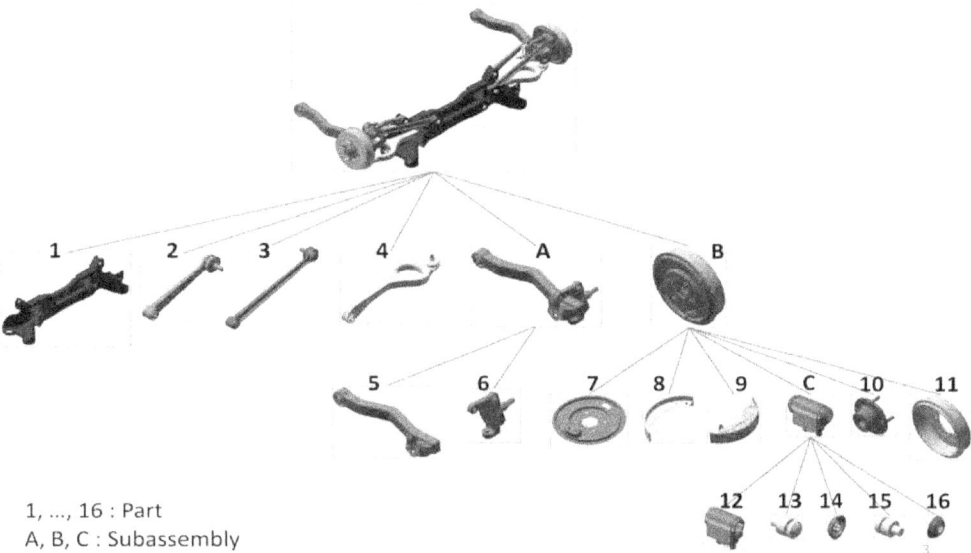

1, ..., 16 : Part
A, B, C : Subassembly

Figure 30. Assembly hierarchy of car suspension module

The connection graph for the car suspension module is shown in Figure 31. The nodes indicate the parts, and the edges indicate the connection relationships between two parts. The red dotted rectangles indicate the ranges of sub-assemblies. A connection graph can be used as an input for a disassembly sequence planning system.

Figure 32 shows the instance diagram of the disassembly information model, which contains the assembly hierarchy of the car suspension module. The instance diagram has a hierarchical relationship between assemblies and parts as well as connection relationships between parts.

65

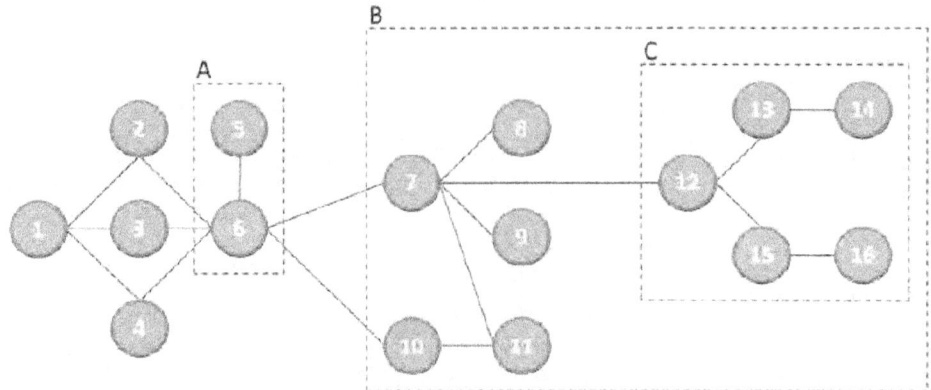

Figure 31. Connection graph

Figure 33 shows an example of a more detailed connection relation between two parts. Parts 10 and 11 have two connections between them, namely one *pin-hole* connection and four *bolt-nut* connections. It is noted that the *pin-hole* connection is established just by two assembly features, while the *bolt-nut* connections are established by a connector applied on two assembly features. In this case, the *pin-hole* connections are established by putting the *pin* feature of part 10 into the *hole* feature of part 11, but the *bolt-nut* connection is established by applying the *bolt-nut* connector to the *hole* features of parts 10 and 11.

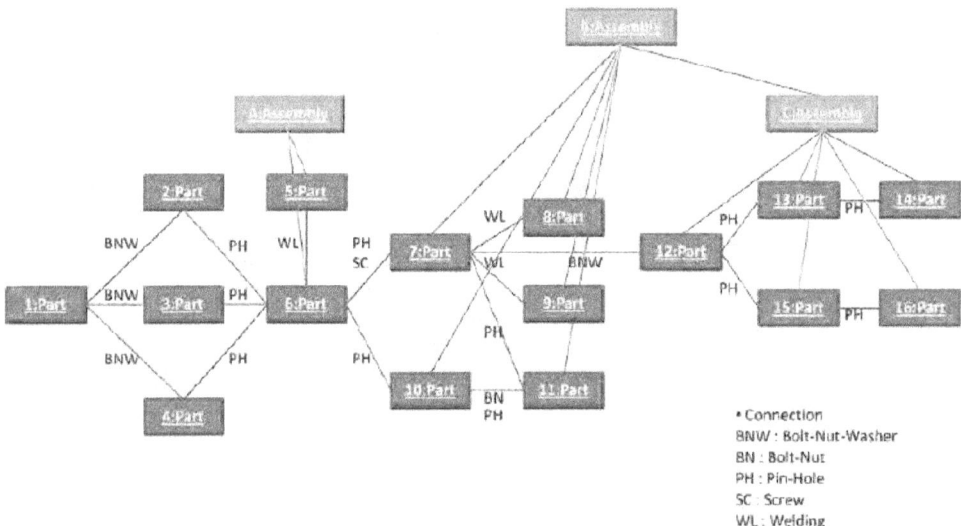

Figure 32. Instance diagram of the car suspension module assembly

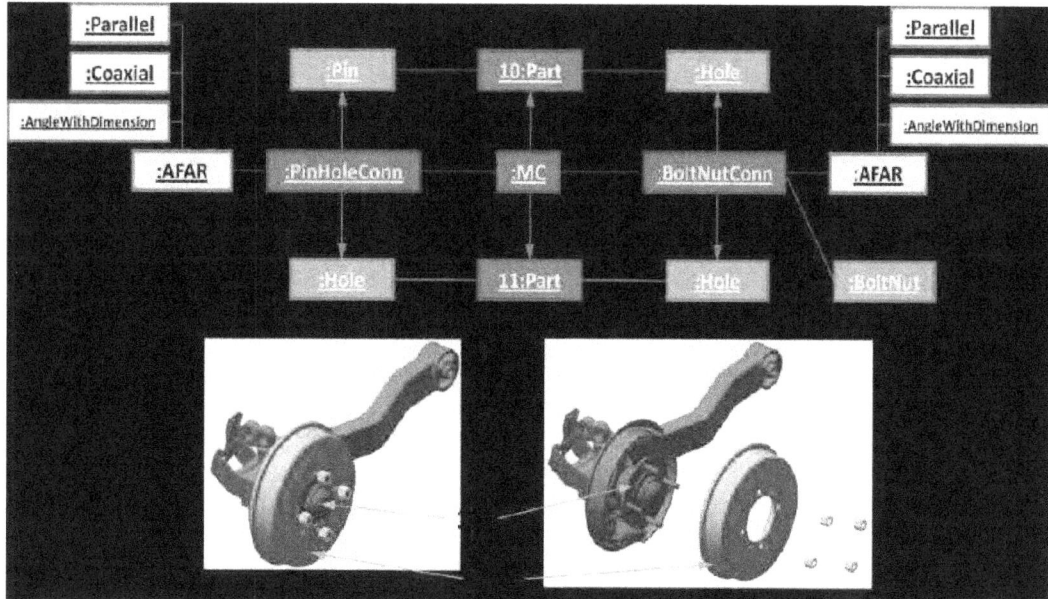

Figure 33. Instance diagram of a sub-assembly (detailed)

A disassembly process planning system uses the connection graph as input and generates a disassembly sequence as shown in Figure 34. The car suspension module can be disassembled through three levels of disassembly sequences

Figure 35 shows an example of the representation of a disassembly sequence using the disassembly information model. A project is composed of several processes, each process has several sub-processes. For disassembly, a sub-process can be a non-destructive or destructive separation. A separation class has an assembly as input and two sub-assemblies as output, and it needs several tasks to separate the sub-assemblies from the input assembly. Each task class has a corresponding connection class of two sub-assemblies.

Figure 36 shows relation between the instance diagrams of the disassembly information model representing disassembly sequences and connection graph. NDSeparation means non-destructive separation. It is an operation, and its class NonDestructiveSeparation is described in Section 3.6.2. The *NDSeparation* is an operation that separates the assembly into the subassembly consisting of parts *1*, *2*, *3*, and *4* and the subassembly consisting of assembly *A* and *B*. There are three *pin-hole* connections between two subassemblies, so this *NDSeparation* has three *pull* tasks to disconnect the *pin-hole* connections.

4.2 Flip-top cell phone

The flip-top cell phone is assumed to be composed of thirteen parts, as shown in Figure 37.

The connection graph for the cell phone is shown in Figure 38. The nodes indicate the parts and the edges indicate the connection relationships between two parts. A connection graph can be used as an input for a disassembly sequence planning system.

Figure 39 shows the instance diagram of the disassembly information model, which contains the assembly hierarchy of the cell phone. The instance diagram has connection relationships between parts. For example, there is a *screw* connection between parts 2 and 6.

A disassembly process planning system uses the connection graph as input and generates a disassembly sequence as shown in Figure 40.

Figure 41 shows an example of the representation of a disassembly sequence using the disassembly information model. To separate the assembly into the subassembly (1,2,3,4,5,6) and the subassembly (7,8,9,10,11,12,13), the pin-hole connection between the subassemblies should be eliminated. The *NDSeparation* has a *pull* task to do this.

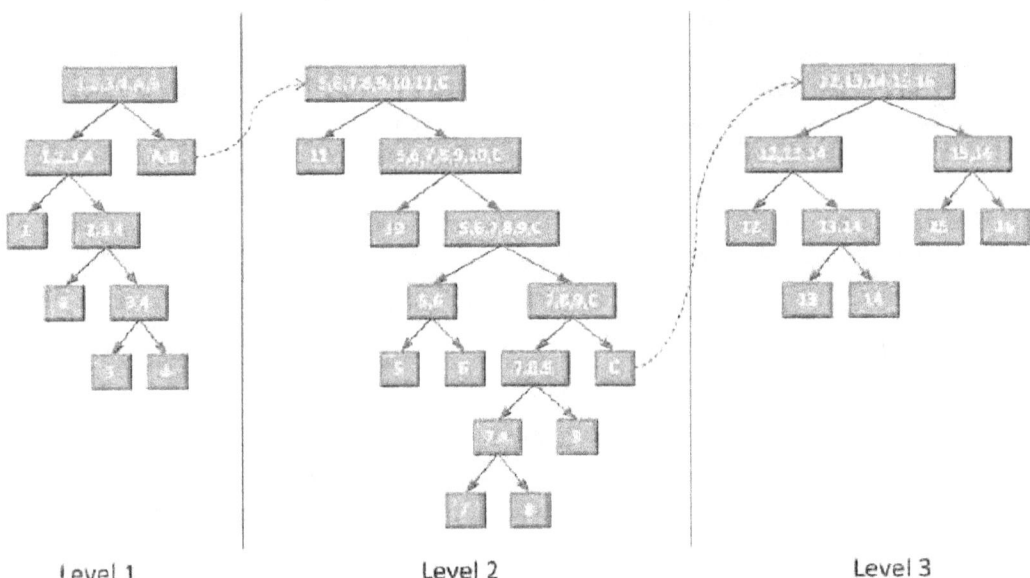

Level 1   Level 2   Level 3

Figure 34. Disassembly sequence of the car suspension module

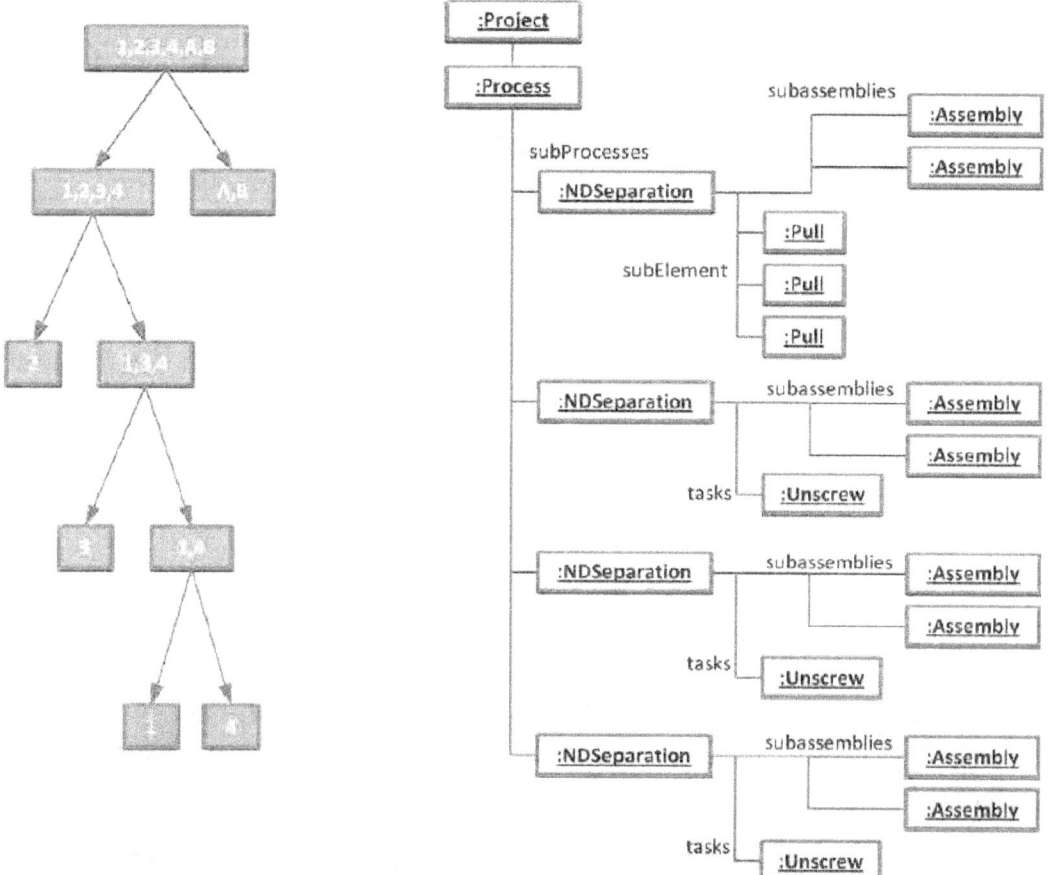

Figure 35. Relation between disassembly sequence and connection graph

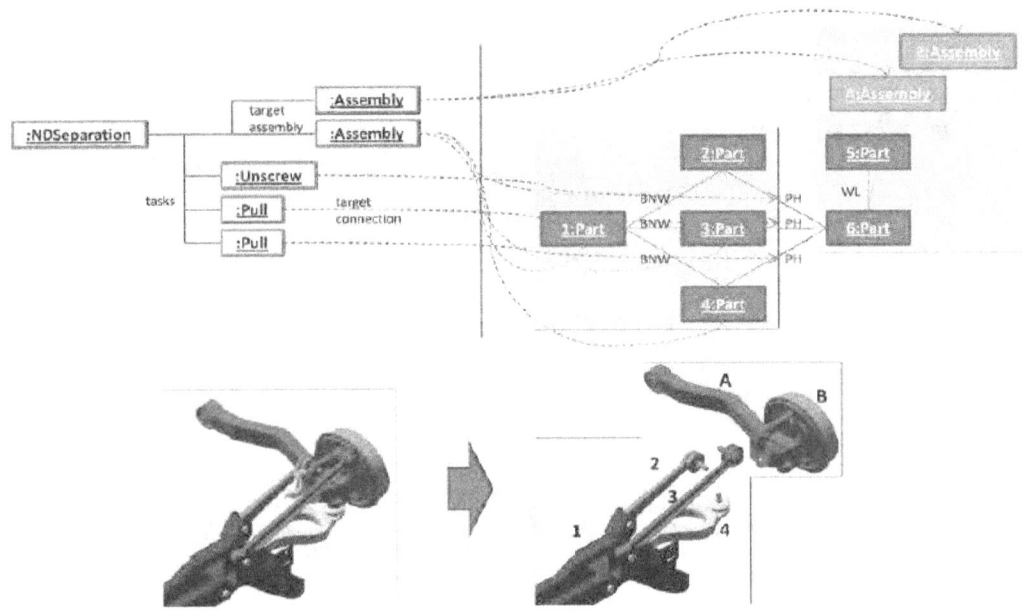

Figure 36. Relation between disassembly sequence and connection graph

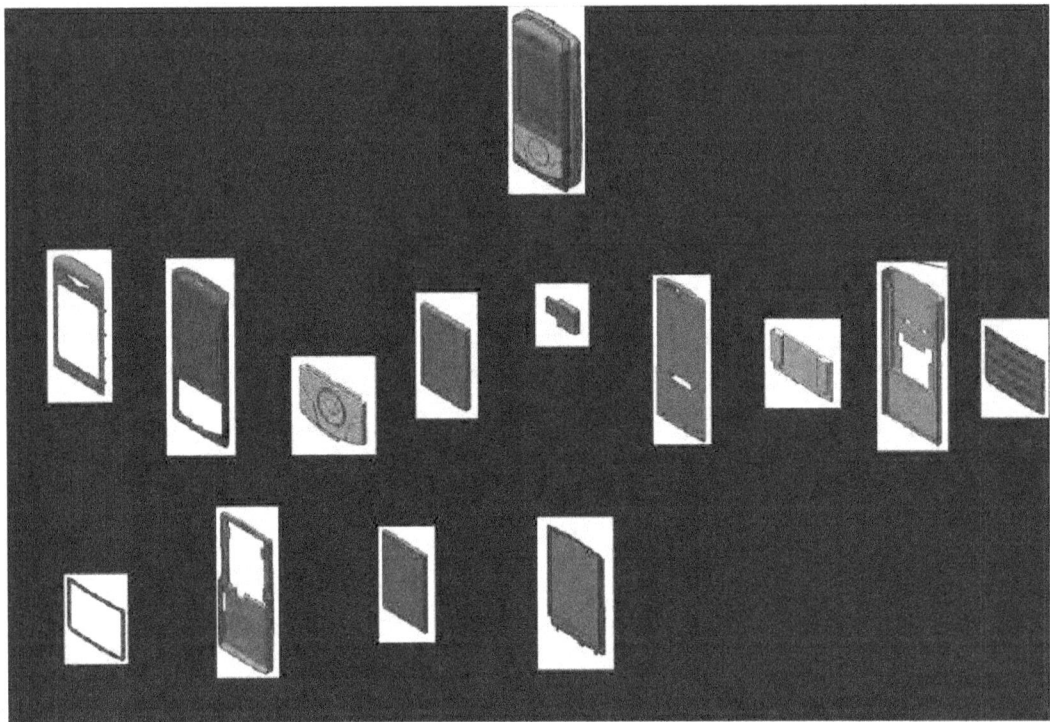

Figure 37. Assembly sequence of flip-top cell phone in a tree structure

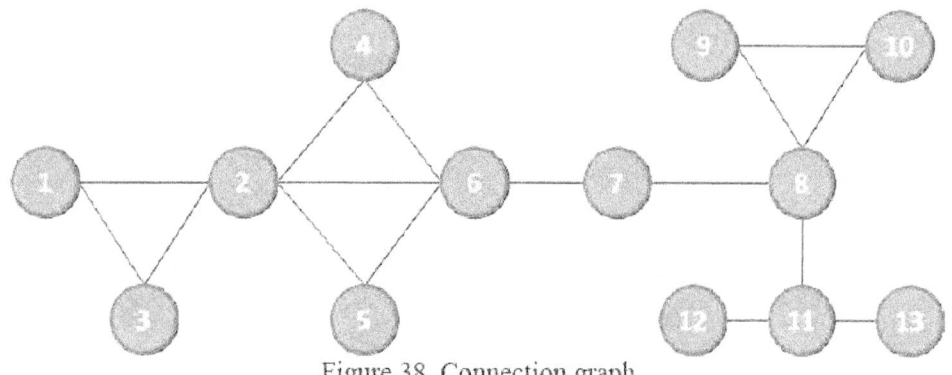

Figure 38. Connection graph

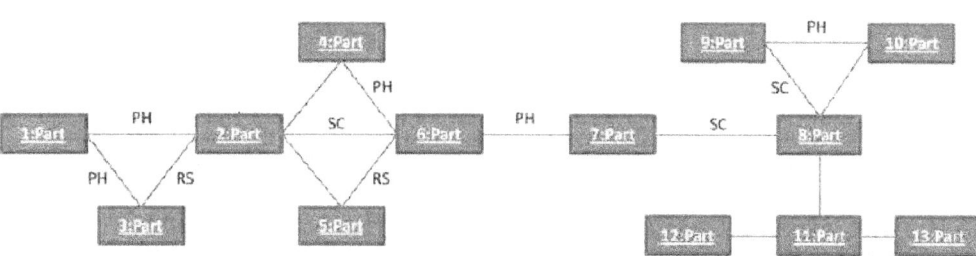

• Connection
PH : Pin-Hole
SC : Screw
RS : Rib-Slot

Figure 39. Instance diagram of a cell phone

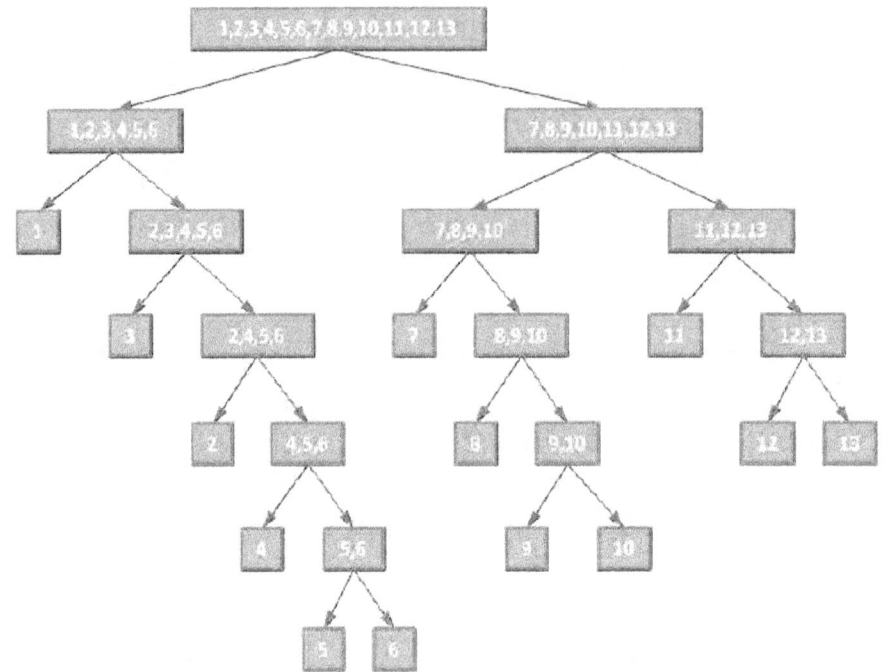

Figure 40. Disassembly sequence of a cell phone

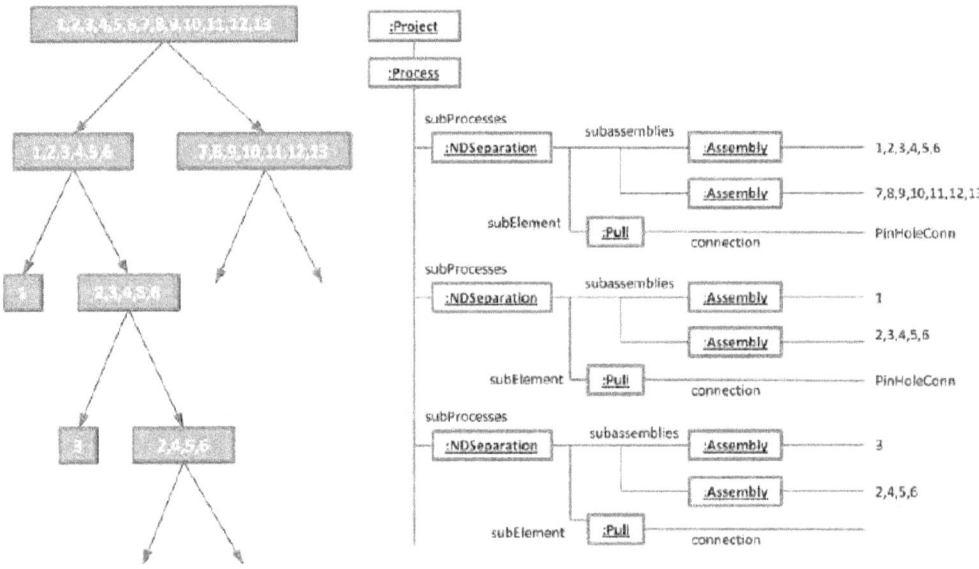

Figure 41. Instance diagram of the disassembly sequence of a cell phone

5. Conclusions and Future Work

Manufacturing industries are facing the challenge of reuse and recycle of products at the end of their service lives. Our literature review shows that the number of environmental protection regulations is increasing rapidly. Reuse and recycle are critical activities to alleviate natural resource depletion and save energy to achieve the goal of sustainable development. Disassembly of end-of-service-life products is a key operation to separate the product into reusable and recyclable parts. Information on design for disassembly and the disassembly process is critical for decision making in design and manufacturing. An information model for disassembly is, hence, developed using the Unified Modeling Language. The model includes all the classes and their relationships on disassembly features, tolerances, dimensions, destructive disassembly, nondestructive disassembly, disassembly methods, equipment, and disassembly workflow. A study of two cases, a vehicle suspension and a flip-top mobile phone, is conducted to initially test the validity of the model. Specifically, information classes of part and feature relationships of parts or subassemblies to be separated, disassembly sequence, disassembly processes have been tested. The other portion of the model is based on available standards.

Possible future work includes more comprehensive tests with more complicated designs. These are needed to test the model. Prototype disassembly database, cost of disassembly analysis software, and disassembly process planning systems can be developed using the information model. Finally, a standard data exchange format of the UML model for design for disassembly and disassembly process plans can also be developed.

Acknowledgements

The authors thank Gwangsub Chang of Ajou Universiy for providing CAD models for cases study, Samer Gafour of the University of Claude Bernard Lyon 1 for collaboration on form feature development, and Tina Lee for many suggestions on improving the quality of the report.

References

[Akagi 1980] Akagi, F., Osaki, H., and Kikiuchi, S., "The method of analysis of assembly work based on the fastener method," Bull JSME, Vol.23, No.184, pp. 1670 - 1675, 1980.

[B89 1985] ANSI/ASME B89.1.12M, Methods for Performance Evaluation of Coordinate Measuring Machines, The American Society of Mechanical Engineers, New York City, New York, 1985.

[B89 2008] B89.4.10360.2, Acceptance test and reverification test for coordinate measuring machines (CMMs), Part 2: CMMs used for measuring linear dimensions. The American Society of Mechanical Engineers, New York City, New York, 2008.

[Bogue 2007] Bogue, R., "Design for Disassembly: A Critical Twenty-first Century Discipline," Assembly Automation, Volume 27, No. 4, pp. 285 - 289, 2007.

[Chan 2003] Chan, C. and Tan, S., "Generating assembly features onto split solid models," Computer-Aided Design, Vol.35, pp.1315 - 1336, 2003.

[De Fazio 1993] De Fazio, T. and Whitney, D., "A prototype of feature-based design for assembly," Journal of Mechanical Design, Vol.115, pp. 723 - 734, 1993.

[Deneux 1999] Deneux, D., "Introduction to assembly features: an illustrated synthesis methodology," Journal of Intelligent Manufacturing, Vol.10, pp. 29 - 39, 1999.

[Desai 2003] Desai, A. and Mital, A., "Evaluation of Disassemblability to Enable Design for Disassembly in Mass Production," International Journal of Industrial Ergonomics, Vol. 32, pp. 265 – 281, 2003.

[Desai 2005] Desai, A. and Mital, A., "Incorporating Work Factors in Design for Disassembly in Product Design," Journal of Manufacturing Technology Management, Vol. 16, No. 7, pp. 712 - 732, 2005.

[Fenves 2002] Fenves, S. J., "A Core Product Model For Representing Design Information," National Institute of Standards and Technology, NISTIR 6736, October 2002.

[Green 2007] Green, J., Aluminum Recycling and Processing for Energy Conservation and Sustainability, ASM International, 2007.

[Gui 1994] Gui, J. K., Mantyla, M., "Functional understanding of assembly modeling," Computer-Aided Design, Vol.26, No.6, pp. 435 - 451, 1994.

[Hamidullah 2006] Hamidullah, Bohez, E. and Irfan, M. A., "Assembly features: definition, classification, and instantiation," IEEE-ICET 2006 2nd International Conference on Emerging Technologies, Preshawar, Pakistan, pp.617-623, November 2006.

[Holland 2000] Holland, W. V. and Bronsvoort W. F., "Assembly features in modeling and planning," Robotics and Computer Integrated Manufacturing, Vol.16, pp. 277 - 294, 2000.

[Homem 1990] Homem de Mello, L.S. and Sanderson, A. C., "AND/OR graph representation of assembly plans," IEEE Transactions on Robotics and Automation, Vol.6, No.2, pp.188 - 100, 1990.

[ICSG 2006-1] ICSG Press Release, Forecast 2008 – 2009, International Copper Study Group, 8 October 2008.

[ICSG 2006-2] International Copper Study Group (ICSG), ICSG Insight - ICSG Secretariat Briefing Paper, No. 1, December 2006.

[Ijomah 2007] Ijomah, W., McMahon, C., Hammond, G., and Newman, S., "Development of Design for Remanufacturing Guidelines to Support Sustainable Manufacturing," Journal of Robotics and Computer-Integrated Manufacturing, Vol. 23, pp. 712 – 719, 2007.

[ISO 8601] ISO 8601, Data elements and interchange formats - Information interchange - Representation of dates and times, International Organization of Standardization, Geneva, Switzerland, 1988.

[ISO10360 2000] ISO 10360-1, Geometrical Product Specifications – Acceptance and reverification tests for coordinate measuring machines, Part 1: Vocabulary, International Organization of Standardization, Geneva, Switzerland, 2000.

[ISO10303 1994-1] Kennicott, P. R., "ISO TC 184/SC4: Product Data Representation and Exchange, Part: 44, Title: Industrial Automation Systems and Integration Product Data Representation and Exchange – Integrated Generic Resources: Product Structure

Configuration (November 1994)," International Organization for Standardization, Geneva, Switzerland, 1994

[ISO10303 1994-2] ISO 10303 Part 203, Industrial automation systems and integration – Product data representation and exchange – Part 203: Application protocol: Configuration Controlled 3D Designs of Mechanical Parts and Assemblies, International Organization for Standardization, Geneva, Switzerland, 1994.

[ISO10303 2000] Sugimura, N. and Ohtaka, A., "ISO TC 184/SC4/WG12 N597, JNC Proposal of STEP Assembly Model for Products (June 2000)," ISO, 2000.

[ISO10303 2001] ISO 10303 Part 224, Industrial automation systems and integration - Product data representation and exchange - Part 224: Application protocol: Mechanical product definition for process planning using machining features, International Organization for Standardization, Geneva, Switzerland, 2001.

[ISO10303 2007] ISO 10303 Part 111, Industrial automation systems and integration - Product data representation and exchange - Part 111: Integrated application resource: Elements for the procedural modelling of solid shapes, International Organization for Standardization, Geneva, Switzerland, 2007.

[ISO 1302 2002] ISO 1302, Geometrical Product Specifications (GPS) - Indication of surface texture in technical product documentation, International Organization for Standardization, Geneva, Switzerland, 2002.

[Jawahir 2008] M'Saoubi, R., Outeiro, J., Chandrasenkaran, H., Dillon O., and Jawahir, I., "A review of surface integrity in machining and its impact on functional performance and life of machined products," International Journal of Sustainable Manufacturing, Volume 1, Nos. 1/2, pp. 203 - 236, 2008.

[Jofre 2005] Jofre, S. and Morioka, T., "Waste Management of Electric and Electronic Equipment: Comparative Analysis of End-of-Life Strategies," Journal of Material Cycles and Waste Management, Volume 7, pp. 24 - 32, 2005.

[Jovane 2008] Jovane, F., Yoshikawa, H., Alting, L., Boer, C., Westkamper, E., Williams, D., Tseng, M., Seliger, G., and Paci, A., "The incoming global technical and industrial revolution towards competitive sustainable manufacturing," CIRP Annals – Manufacturing Technology, Vol. 57, pp. 641 - 659, 2008.

[Kroll 1998] Kroll, E. and Hanft, T., "Quantitative Evalution of Product Disassembly for Recycling," Journal of Research in Design, Vol. 10, pp. 1 - 14, 1998.

[Kumar 2008] Kumar, V. and Sutherland, J., "Sustainability of the Automotive Recycling Infrastructure: Review of Current Research and Identification of Future Challenges," International Journal of Sustainable Manufacturing, Volume 1, Nos. 1/2, pp. 145 - 167, 2008.

[Kuo 2000] Kuo, T., Zhang, H., and Huang, S., "Disassembly analysis for electromechanical products: a graph-based heuristic approach," International Journal of production research, Vol.38, No.5, pp. 993 - 1007, 2000.

[Lambert 1997] Lambert, A., "Optimal disassembly of complex products," International Journal of Production Research, Vol.35, No.9, pp. 2509 - 2523, 1997.

[Lambert 2003] Lambert, A., "Disassembly sequencing: a survey," International Journal of Production Research, Vol.41, No.16, pp. 3721 - 3759, 2003.

[Lambert 2005] Lambert, A. and Gupta, S., "Disassembly Modeling for Assembly, Maintenance, Reuse, and Recycling," CRC Press, 2005.

[Lee 1985] Lee, K. and Andrews, G., "Inference of positions of components in an assembly: part 2," Computer Aided Design, Vol.17, No.1, pp.20 - 24, 1985.

[Lee 1996] Lee, K. and Gadh, R., "Computer Aided Design for Disassembly: A Destructive Approach," Proceedings of the 1996 IEEE International Symposium on Electronics and the Environment, pp. 173 - 178, May 1996.

[Li 2002] Li, J., Khoo, L.. and Tor, S., "A novel representation scheme for disassembly sequence planning," International Journal of Manufacturing Technology, Vol.20, pp. 621- 630, 2002.

[Li 2005] Li, J.R. Khoo, L,P. and Tor, S.B., "An object-oriented intelligent disassembly sequence planner for maintenance," Computers in Industry, Vol.56, pp. 699 - 718, 2005.

[Moore 1998] Moore, K., Gungor, R. and Gupta, S., "Disassembly process planning using Petri nets," Proceedings of 1998 IEEE Conference on Electronics and the Environment, pp. 88 - 93, 1998.

[Nasr 2006] Nasr, N. and Thurston, M., "Remanufacturing: A Key Enabler to Sustainable Product Systems," Proceedings of the 13th CIRP International Conference on Life Cycle Engineering, Leuven, Belgium, pp. 15 - 18, 2006.

[REACH 2006] Registration, Evaluation, Authorisation and Restriction of Chemical substances (REACH), EC 1907/2006, European Community, 2006.

[Rifer 2009] Rifer, W., Brody-Heine, P., Peters, A., and Linnell, J., "Closing the Loop Electronics Design to Enhance Reuse/Recycling Value," the Green Electronics Council, Portland, Oregon, January 2009.

[RoHS 2002] Restriction of Hazardous Substances Directive (RoHS), the directive on the restriction of the use of certain hazardous substances in electrical and electronic equipment, 2002/95/EC, European Community, 2002.

[Rumbaugh 1999] Rumbaugh, J., Jacobson, I., and Booch, G., The Unified Modeling Language Reference Manual, Addison Wesley, 1999.

[Seliger 2007] Seliger, G., "Sustainability in manufacturing: recovery of resources in product and material cycles," Springer, 2007.

[Shah 1992] Shah, J. and Tadepalli, R., "Feature based assembly modelling", Proceedings of 1992 ASME Computers in Engineering Conference and Exposition, San Francisco, CA, USA, pp. 253 - 260, August 1992.

[Soldhi 1991] Soldhi, R. and Turner, J., "Representing tolerance and assembly information in a feature-based design environment," ASME Advanced Design Automation, pp.101-108, 1991.

[Sriram 2000] Sriram, R., Navinchandra, D., and Allen, R., Environmental Issues in Collaborative Design, Mechanical Life Cycle Handbook: Good Environmental Design and Manufacturing, Hundal, M. (editor), Mercel Dekker , Inc, 2000.

[Sudarsan 2003] Sudarsan, R., Han, Y., Feng, S., Roy, U., Wang, F., Sriram, R., and Lyons, K., "Object Oriented Representation of Electro-Mechanical Assemblies Using UML," National Institute of Standards and Technology, NISTIR 7057, October 2003.

[Tang 2000-1] Tang, Y., Zhou, M. and Caudill, R., "An integrated approach to disassembly planning and demanufacturing operation," Proceedings of 2000 IEEE International Symposium on Electronics and the Environment, pp. 354 - 359, 2000.

[Tang 2000-2] Tang, Y., Zhou, M., Zussman, E., and Caudill, R., "Disassembly Modeling, Planning, and Application: A Review," Proceedings of the 2000 IEEE

International Conference on Robotics & Automation, San Francisco, CA, April 2000, pp. 2197 – 2202.

[Tseng 1999] Tseng, H. and Li, R., "A novel means of generating assembly sequences using the connector concept," Journal of Intelligent Manufacturing, Vol.10, No.5, pp. 423 - 435, 1999.

[WEEE 2002] Waste Electrical and Electronic Equipment Directive (WEEE), 2002/96/EC, European Community, 2002.

[Y14.5 1994] ANSI/ASME Y14.5M, Dimensioning and Tolerancing, The American Society of Mechanical Engineers, New York City, New York, 1994.

[Yin 2003] Yin, Z., Ding, H., Li, H., and Xiong, Y., "A connector-based hierarchical approach to assembly sequence planning for mechanical assemblies," Computer-Aided Design, Vol.35, pp. 37 - 56, 2003.

[Zha 2002] Zha, X. and Du, H., "A PDES/STEP-based model and system for concurrent integrated design and assembly planning," Computer-Aided Design, Vol.34, pp.1087 - 1110, 2002.

[Zussman 2000] Zussman, E. and Zhou, M., "Design and implementation of an adaptive process planner for disassembly processes," IEEE Transactions on Robotics and Automation, Vol.16, No.2, pp. 171 - 179, 2000.